D0781056

Engaging Geopolitics

Engaging Geopolitics

Kathleen E Braden
Seattle Pacific University

Fred M Shelley
Southwest Texas State University

PRENTICE HALL

An imprint of PEARSON EDUCATION

Harlow, England · London · New York · Reading, Massachusetts · San Francisco · Toronto · Don Mills, Ontario · Sydney
Tokyo · Singapore · Hong Kong · Seoul · Taipei · Cape Town · Madrid · Mexico City · Amsterdam · Munich · Paris · Milan

Pearson Education Limited
Edinburgh Gate, Harlow
Essex CM20 2JE
England

and Associated Companies throughout the World

Visit us on the World Wide Web at: http://www.awl-he.com

© Pearson Education Limited 2000

The right of Kathleen E. Braden and Fred M. Shelley to be identified as authors of this Work has been asserted by them in accordance with the Copyright, Designs and Patents Acts 1988.

All rights reserved. No part of this publication may be reproduced, stored in a retrieval system, or transmitted in any form or by any means, electronic, mechanical, photocopying, recording, or otherwise without either the prior written permission of the Publishers or a licence permitting restricted copying in the United Kingdom issued by the Copyright Licensing Agency Ltd, 90 Tottenham Court Road, London W1P 0LP.

First published 2000

ISBN 0 582 03565 1

British Library Cataloguing-in-Publication Data
A catalogue record for this book is available from the British Library

Library of Congress Cataloging-in-Publication Data
A catalog record for this book is available from the Library of Congress

Typeset by 3 in Sabon and News Gothic.

Printed and bound by Addison Wesley Longman, Singapore (Pte) Ltd. Printed in Singapore

Contents

Acknowledgements

We are grateful to the following for permission to reproduce copyright material;

Pearson Education Ltd for figure 2.1 from AG Hopkins *An Economic History of West Africa* (Longman, 1973), figure 2.4 from Klaus Dodds *Geopolitics in a Changing World* (Prentice Hall, 1999), figures 3.1a–d from M Chamberlain *The Longman Companion to European Decolonisation in the Twentieth Century* (Addison Wesley Longman, 1998), figure 4.4 from Robert B Potter et al *Geographies of Development* (Addison Wesley Longman, 1999); John Wiley & Sons for figure 6.2 from M Glassner *Political Geography* (John Wiley & Sons, 1993); US News & World Report and Dale Glasgow for figure 5.4 from the May 9, 1988 issue of *US News & World Report*

Whilst every effort has been made to trace owners of copyright material, in a few cases this has proved impossible, and we would like to take this opportunity to apologise to any copyright holders whose rights we may have unwittingly infringed.

1

Introduction

The word "geopolitics" is familiar to most people, yet difficult to define. It has an aura of the military strategist's world – theoretical, manipulative, moving game pieces across a giant world map. Few subdisciplines of the social science we call geography may seem farther removed from the ordinary facets of human behavior than the realm of geopolitics.

Fred Shelley and I have created this book with the goals of making geopolitics more accessible to the beginning students and of drawing the discipline's concepts nearer to questions relevant to ordinary citizens.

Few fail to notice that the world map and international political relations are changing rapidly at the end of the twentieth century, but a coherent paradigm for organizing that change seems elusive. The superpower blocs that dominated the geopolitical scene after World War II have given way to a mystery. Is a "new geopolitics" indeed being born or are we witnessing merely business as usual with world political geography? And if we cannot firmly pin down how to analyze a system in development, how can we convince the novice observer that geopolitics is relevant to his or her well-being?

1.1 Beginning in Central Asia

One example of this relevance may be suggested by the case of Kazakhstan, where I taught geopolitics to students at Almatinsky University just a few years after the break-up of the Soviet Union. My students (a mixture of young Kazakhs, Russians, Jews, Uighurs, and Kurds) were an enthusiastic group, but uneducated in thinking about political geography because the Soviet system had largely buried the field. I joined them in considering what their perspective on the geopolitical situation of their new country would be – what would it mean for their personal futures? Much of their very security and identity would depend on the concepts Fred and I explore in this book.

To begin with, they were citizens of a newly independent state, now with its own sovereignty or control over both internal and external affairs. This sovereignty would affect their lives in many ways, from the language they used in the workplace to the fact that they would now be crossing an international border to visit cousins in Kyrgyzstan to the south (also a former Soviet republic). They were receiving documents from their new state, and, if they were of mixed ethnic heritage, they might need to make a choice between a Russian or a Kazakh passport and a decision over whether to relocate to Russia under refugee status. In fact, their very national identity was now being formed with each new law passed in Almaty or Moscow. The revival of Kazakh culture allowed some to reach back in time and connect with their historic identity and develop a new sense of territoriality about their homeland. But the borders of this territory were very much an open question: the restlessness of Russian populations along the border with Siberia was inspiring the Kazakh government to move the capital city northward to establish a firmer hold on the region. Even the name of this new capital showed the geopolitical forces at work in their lives: the students who grew up calling the city Tselinograd were now referring to it as Aqmola. The border to the east with China was under dispute, and radiation from Chinese nuclear tests perceived to be wafting across that permeable barrier was a source of fear for the local people.

Not all the change was comfortable for the students and internal divisions became apparent to me. National identity and state formation required renewed definitions of "outsider" and "insider". The Uighur and Russian students seemed uneasy and separated from their Kazakh colleagues. It may have been so under the Soviet system as well, but now political change threw a spotlight on these differences.

Perhaps most of all, Kazakhstan's position in the capitalist world structure was being formulated, and the students exhibited an aura of excitement and anticipation about their futures. They would study "business", learn English, and buy a good automobile. They would live in an open society, travel, and take their place in a modern, global culture. Their new state was reaching out to form economic alliances with Europe, the United States, Turkey, and the Middle East. As they shook off the remnants of Russian control, new cultural hegemonies seemed to influence them more than they knew, from their favorite MTV show to the Snickers candy bars they brought to class and to the fact that most wanted to study English more eagerly than to learn their native Kazakh. I watched them struggle with the irony of a nationalist revival that was already married to a globalization of their culture.

In short, the new state of Kazakhstan affected my students' security, profession, environment, and even sense of self. To them, geopolitics could not have been more personal.

1.2 Goals of the authors and the meaning of our words

We conceived of this textbook before the devolution of the USSR and originally intended to make a connection between geopolitical analysis and ethical choices. Not interested in a morally neutral treatment of US–Soviet relations, we were both horrified at the nuclear build-up occurring and hoped to offset the discourse that Simon Dalby explores so well in his critique of the Reagan-era Committee on the Present Danger. In fact, Dalby warns against a traditional geopolitics that pretends scientific objectivity with "the sort of grand theorizing that provides blueprints and policy advice to foreign policy specialists and strategic thinkers" but, in fact, serves to promote a militarization of domestic politics (Dalby 1990: 180). With the break-up of the USSR and a diminishment of the nuclear war threat (at least in public debate if not in reality), the purpose of our book evolved. We were also daunted by the prospect of doing justice to the very complex field of both theoretical and applied ethics.

But our original purpose of making geopolitics accessible to the lay reader has remained. The very definition of what constitutes the field has been problematic. In the years since Fred and I began to write the textbook, many new and "post" terms of discourse have emerged: new geopolitics, a new century, a new world order, global change, post-Soviet, post Cold War, post-modern. The idea of change is in the air, almost as though we define our era of world relations as a tautology: it is called a new era because it is new. Something is behind us, we are now "post", but what is this new world we are entering? What defines the change? Has anything indeed really changed?

Geopolitical scholars who have asked such questions seem to center on the role of the state as the primary actor in geopolitical relations and have inquired whether this fact is shifting. Is the state still of such importance? In some sense, the new geopolitics of a diminished state role harkens back to an old geopolitics of pre-Enlightenment Europe, when other forms of spatial organization dominated human life (empires, blocs, kingdoms, city-states).

Some scholars have cautioned that the state is still alive and well, very much with us as a force for geopolitical organization of the modern world. Others have argued that the role of geography needs to be revived in international relations – that worries over determinism have caused us to underestimate the role of geography (Harvey Starr in Ward 1992). Many of these questions have been considered in the February 1992 issue of *The*

Professional Geographer in a panel discussion, "The Political Geography of the Post Cold War World". In his comments, Nigel Thrift (p. 6) notes:

> It seems to me that the experiences of the 1980s underline the need increasingly to call into question the nation-state as the basic, organizing principal of political life. We should not go too far. The state still retains very considerable power over structures of security, production, and knowledge (witness the Gulf War). But, in Jessop's memorable phrase, the nation-state is gradually being "hollowed out". The result is that anyone who believes the geography of the post Cold War world is able to be considered simply in terms of remapping a set of nation states is swimming against the tide of history.

In attempting to define a new geopolitics of the post Cold War period, we must bear in mind that geography is still the parent discipline of geopolitics. To understand what is basic to geopolitical inquiry, one must also understand what is basic to all geographic thought and what distinguishes the discipline from the other social sciences: issues of place, location, scale, region, boundaries, and above all, a spatial perspective on human behavior. If the significance of geography eludes us, then geopolitics becomes a mere exercise in international relations. The context for human action in the world (still largely expressed through a system of states and interstate relations) must be understood as set in both time and space. Taylor, Watts, and Johnston review what they term "multiple geographies" in terms of time and space as containers for human activities and suggest that the fusion of the long-term perspective of time with the global scale of space may be helpful in making sense of global change (Johnston *et al.* 1995).

The tools in our arsenal include mainly the map and the written word. Maps force spatial expression and language forces terms by which we understand relations. Simon Dalby is among the many who have explored social theory as it relates to the politics of domination, the very role of language in creating the inside and outside that then may be expressed spatially on a map. In Chapter 2 of *Creating the Second Cold War: the Discourse of Politics* (Dalby 1990), he extends the notion of "otherness" to territory: theirs and ours. The outsider is barbarian, immoral. Classic geopolitical theoreticians such as Mackinder and Spykman have been used to bolster such thinking. The words chosen create the perception of reality; the map can then be manipulated to reinforce those notions. The Russians of northern Kazakhstan are "the others", but it is not merely the location of the new capital on the map that is important; the very word used for its name suggests inclusion or exclusion in both historic and geographic context.

Dalby calls us to a critical geopolitics that eschews the promise of a truly neutral position, but invites consideration of power relations and how our discourse can actually contribute to those relationships. Our textbook cannot aspire to break ground in the geopolitics that is Dalby's vision, but we do try to engage the student to demythologize geopolitics, convince the reader that responsibility to understand geopolitical relationships may be a valuable first step to entering into the discourse, rather than giving up the ground to a falsely dispassionate treatment of state security. The authors' language in this text reveals our agenda.

1.3 Questions for the reader

If what is "new" in geopolitics remains elusive, it may help to express questions that arise in the course of this book. The list below forms a constant thread throughout most of the chapters.

1. Is the state gaining or losing importance as the prime unit of geopolitical analysis?
2. Is the concept of sovereignty eroding or intact? How do we understand a world in which expression of self-determination is held in high regard along with the desire to keep states intact? How can this apparent contradiction of aims be reconciled?
3. Are social processes occurring that are different from past historic periods? (This question is especially relevant to the consideration of nationalism.)
4. Do boundaries still matter? A state must above all have territory – it can be seen on a map. To define territory, boundaries must be drawn.

Plate 1.1 Border between Egypt and Israel at the Sinai–Gaza Strip in the mid-1980s (photo: Kathleen Braden)

These lines are by their nature exclusionary, but to serve the purpose of defining "otherness", they must be intact or at least perceived to be intact. Technological changes in warfare, industrial processes, ability to conduct criminal activities, transfer of information and culture, and increasing diffusion of harmful environmental impacts all speak to the notion that boundaries are quite permeable. Must a boundary be impermeable to be intact?

5 What is the role of scale (from local to regional to global)? Johnston *et al.* (1995: 378) warn about turning globalization into a "superficial slogan" and note: "global change does not in any sense make other geographical scales disappear, quite the reverse in fact: the rise of 'globalization' coincides with a simultaneous affirmation of 'localization' as places both of control (e.g. world cities) and of resistance (e.g. new nationalisms)." But the question of how local is "local" in today's world seems very relevant. If technological shift has made borders less impermeable, then the ability to exclude and therefore establish "us" or "local" becomes more problematic.

6 Can a set of transboundary moral codes be established and respected if the near future holds both a continued state system as well as the breakdown of boundaries propelled by technological change?

1.4 Book organization

There are seven main chapters of varying length in the book, each organized with an initial set of keywords and key propositions. Chapters are subdivided, and boxed material often provides supplementary illustrations for the reader. Maps and graphics display spatial relations discussed in the text.

Fred Shelley is the primary author of Chapters 2–4 and Kathleen Braden is the primary author of Chapters 5–9, as well as Chapter 1.

Our hope is that the book will be a starting point for students who will be moving vertically into more advanced courses in political geography or laterally into other concerns of international affairs.

2

Fundamental Concepts of Geopolitics

KEYWORDS

geography,
nation,
state,
sovereignty,
boundaries,
territory,
heartland

KEY PROPOSITIONS

- Geopolitics focuses on the geographical perspective of international relations.
- The distinction between nation and state is crucial to understand and analyze many international conflicts.
- There is a close relationship between power and territory at all geographic scales, sometimes leading to conflict over territorial control.
- Geopolitics as a discipline has been influenced by British, French, German, American, and Russian theories.

Geopolitics is the study of international relations and conflicts from a geographical perspective. The geographical perspective suggests that location, distance, and the distribution of natural and human resources have significant influences on international relations. Thus we begin our investigation of geopolitics by focusing on the unique perspective of the geographer. We then investigate several key geopolitical concepts: the nation, the state, power, territory and conflict.

2.1 The geographical perspective

Geography is the systematic study of location and place. Professional geographers address questions concerning where and why various phenomena are located and distributed. In addition, they examine and compare the unique characteristics of places while considering the relationships between individual places and the global economy.

What distinguishes contemporary geography from other approaches to knowledge is not content, but intellectual approach. None of the vast number of subjects analyzed by professional geographers are unique to the discipline of geography. Geography is distinguished from these disciplines by its unique, holistic and integrative approach to knowledge. The geographer pulls together knowledge of social, economic, political, cultural and environmental forces that shape human activity in places and regions throughout the world.

Geographers address several fundamental questions in their studies of the world around them. These include concepts of location, distance, direction, distribution, diffusion, place and region. Questions of location address perhaps the most fundamental of all geographical questions: where is it?

In examining the locations of places, geographers distinguish between absolute and relative location. Absolute location is the exact position of a place on the surface of the earth, regardless of other places. On a global scale, we use latitude and longitude to determine absolute location. No two places have exactly the same latitude and longitude. At other times, however, geographers are concerned about the location of places with respect to other places. Statements which base knowledge of location on the locations of other places are statements of relative location. As we shall see, relative location is critical to the understanding of international geopolitics. For example, Germany's position at the center of continental Europe, between France and England to the west and Russia and eastern Europe to the east, has proven critical to the development of German geopolitics throughout its history.

Questions of distance and direction, like questions of location, can be conceptualized in absolute or relative terms. Often, relative distance is measured in terms of travel time or cost rather than in miles, kilometers or other spatial units. Knowledge of relative distance is especially critical in a contemporary world characterized by increasing worldwide interdependence, instantaneous telecommunications and reduced travel time between places.

The distribution of a phenomenon is the arrangement of that phenomenon on the surface of the earth, or some part of the earth's surface. Geographers have long expressed concern with the geometric properties of distributions as well as with understanding the processes responsible for creating and changing distributions. Many different types of distributions are critical to the understanding of geopolitics and international relations. These include the distribution of population, natural resources, industries, military expenditures, cultures, ethnic groups and many others.

Distributions are subject to change over time, and changes in distribution can have profound effects on international relations. Diffusion is the process by which new ideas, innovations or technologies are transmitted from one individual, group or country to another across space and over time. For example, the Industrial Revolution diffused from its origin in England in the late eighteenth century to Europe and North America in the nineteenth century, and to most other parts of the world in the twentieth century. Normally, distributions change as a result of diffusion processes. Hence, knowledge of diffusion is often critical to the understanding of changes in spatial distributions. These changes in turn often affect international relations and geopolitics.

Although geographers pay considerable attention to distributions and diffusion processes, they are also quite concerned about the understanding of individual places. For the geographer, a place may be as small as an individual household, or as large as a country or continent. In today's interdependent world, geographers cannot examine the unique qualities or characteristics of individual places without reference to relationships between the place under study and other places. For example, we can hardly expect to understand Europe during the Cold War era without reference to its location between the United States and the Soviet Union. Nor can we understand the geography of Central European cities such as Prague or Vienna without reference to the effects of the Iron Curtain on that city's trade patterns during the Cold War.

Geographers focus not only on individual places but also on regions. A region can be defined as a set of places with common attributes. The common attributes characteristic of regions in the world today can include culture, economy, political system, language, religion and many others. In this book, we will frequently focus on large regions including entire continents or subcontinental areas such as Eastern Europe, East Asia and Latin America. Yet very small regions or communities can be equally important to geographical analysis. What happens in your own community or neighborhood could very well have long-run implications for the entire world.

Geography includes two major sub-disciplines: physical geography and human geography. Physical geographers examine the location and distribution of various aspects of the earth's physical system, including climate, vegetation, geomorphology and ecology. Human geographers, on the other hand, focus on the relationships between human societies and cultures and the earth on which they live. Geopolitics is a subset of human geography. Yet knowledge of the physical geography of particular places is often critical to the understanding of international conflict. For example, Russia's northern location and lack of access to warm-weather ports that can be used for shipping throughout the year has made access to the Mediterranean an important goal of Russian foreign policy for centuries.

Human geographers investigate many different spheres of human activity. Thus urban geographers study the geography of cities; cultural geographers examine the effects of culture and cultural practices on the landscape; and economic geographers investigate locational aspects of economic systems. Political geography is another important sub-field within human geography. Political geographers investigate the relationships between politics and geography at spatial scales ranging from local to international. Concerns of the political geographer include such topics as boundaries and boundary disputes, election procedures and outcomes, land use controversy, law and legal systems, and the management of common resources such as the oceans, Antarctica and outer space.

Geopolitics is that subset of political geography that deals directly with international relations, international conflicts and foreign policies. Of course, students of geopolitics cannot ignore national and local politics in investigating international relations from a geographical perspective. In the United States, for example, the success or failure of foreign policy has often been influenced by domestic political considerations, as in America's refusal to join the League of Nations after World War I or in the mounting opposition to American involvement in Vietnam during the late 1960s. As we shall see, these and many other domestic disputes within countries have had considerable effects on international political conflict.

2.2 Nations and states

The geographic perspective on international relations begins with the fundamental distinction between nations and states. Nations are defined on the basis of culture, religion, language and ethnicity. Nations contain persons who share common cultural traits and a sense of self-identification enabling them to be distinguished from other groups of people living outside the national territory. Examples include the Arabs, Basques, Quebecois, Welsh, Scottish and many others.

A state, on the other hand, is a political unit. In the modern international system, the entire inhabited world is divided into states. Boundaries are drawn in order to separate states. The right of each state to control the territory encompassed by its boundaries is recognized by the international community. Recognition of these territorial rights by the international community is known as sovereignty, and the concept of sovereignty implies recognition within international law that the government of a state has jurisdiction over its territory.

Currently, the world political map includes approximately 200 sovereign states. This number has increased steadily since World War II as a result of the independence of former European colonies in Africa, Asia and Latin America. Since the end of the Cold War, several additional states have been added to the world political map. The break-up of the Soviet Union and Yugoslavia has brought independence to former Soviet and Yugoslav republics such as Lithuania, Armenia and Croatia, while Czechoslovakia has voluntarily divided itself into two states, the Czech Republic and Slovakia.

Although we often use the terms "nation" and "state" interchangeably in ordinary conversation, the distinction between nations and states is crucial to the understanding and analysis of many international conflicts. Russia, France, Germany and the United Kingdom are states, whereas the Russians, the French, the Germans and the English are nations. The boundaries of many states are consistent with the territories inhabited by particular nations. For example, most persons of French nationality live in France, and most Japanese are citizens of Japan. Even these relatively homogeneous states contain minority groups of different nationalities, however, including the Basques and Bretons of France and the Ainu of Japan.

Many other states contain large numbers of nations within the boundaries. Russia, India and Nigeria are all composed of large numbers of nations. Other states, including Canada, Switzerland, Belgium and South Africa, include two or more large nations. On the other hand, many nations are divided among two or more

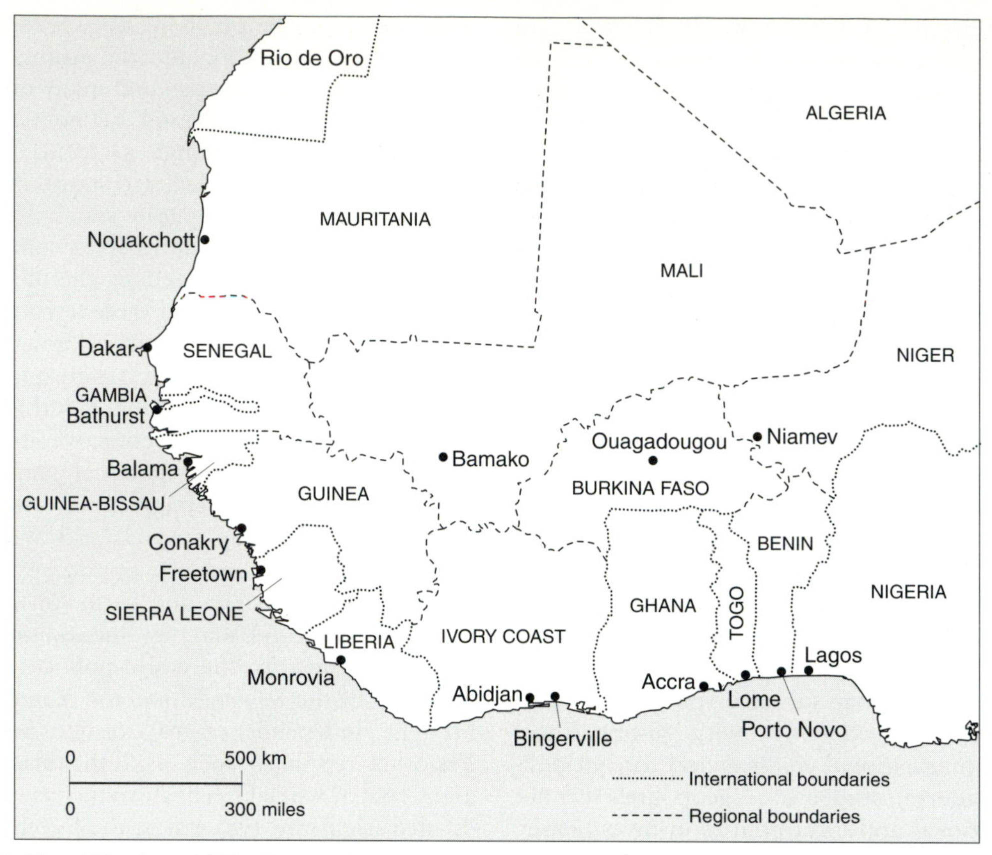

Figure 2.1 West Africa (from AG Hopkins, *An Economic History of West Africa* (Longman, 1973)

states. The Arabs, Kurds, Germans and Koreans are among the many nations of the world that are divided politically among two or more states.

The lack of correspondence between state boundaries and areas of national dominance has resulted in large numbers of international conflicts. Although such conflict is sometimes controlled by formal legal means, in other cases it results in a considerable degree of social disruption, often accompanied by violence. This is especially true when one nation maintains political power and other nations believe that their interests and needs are being suppressed.

In such cases, nationalist movements dedicated to increasing autonomy or political independence of their nations often develop. For example, nationalist movements were critical to the eventual independence of Lithuania, Estonia and other republics of the former Soviet Union. In many other states, distinct nations continue to press for increased autonomy and/or independence. In recent years, such demands have been expressed by Basques in France and Spain, the Kurds in the Middle East, and many others. National minorities within states, such as the Scots and Welsh in the United Kingdom and the Quebecois in Canada, have also expressed demands for independence or greater autonomy.

The lack of correspondence between national territories and state boundaries is an especially significant problem in less developed countries. This failure results from the colonial history of these regions. Boundaries between colonies were drawn at conference tables in European capitals by diplomats with little or no knowledge of local conditions. Boundaries were drawn to suit the convenience of the European colonial masters. Because a primary purpose of colonialism was to provide ready sources of raw materials, and because ocean transportation was the primary

Plate 2.1 Coat of arms of Canadian provinces: Victoria, British Columbia (photo: Kathleen Braden)

means of getting these raw materials to European factories, many colonial boundaries were drawn in such a way as to give each European power access to indigenous port facilities.

This situation is especially evident in West Africa (Figure 2.1). Even a cursory glance at a map of West Africa reveals that the boundaries between most states have been drawn perpendicular to the orientation of the coast, with each country's territory focused on a port city, which during colonial days served as the main gateway between the colony and its mother country. As elsewhere, boundaries were drawn with little or no knowledge of or appreciation for local ecological conditions and cultural considerations. Once the boundaries were delineated, it soon became evident that most cut across national territories. Many indigenous West African cultures had developed in accordance with local ecological conditions. Some were oriented to the tropical rainforest climate of the south, others to the drier savanna further to the north, and others to the semi-arid regions adjacent to the Sahara Desert.

During the late nineteenth and early twentieth centuries, the European colonial powers attempted to integrate residents of their colonies into their European-centered economies. European languages, educational systems, industries, economic activities and political institutions were superimposed on local cultures. As the Europeans integrated their colonies into the global economy, the indigenous people became increasingly torn between national loyalty and obligations to the European-defined colony. Once the former colonies achieved independence, not only were members of different national groups divided in their political loyalties, but each state consisted of a diverse set of national groups, making the achievement of national unity difficult. Tension between competing national groups remains characteristic of many West African countries even today.

2.3 Power and territory

The study of geopolitics involves considerations of territory, power and conflict between nations and states. In international relations, control of territory usually increases power, while increased power can expand control of territory. More powerful states exercise direct or indirect territorial control over weaker ones. In many cases, power implies direct, formal political sovereignty over designated territories. Throughout history, many wars have been fought for control of specific territories, and the foreign policies of many countries have been influenced by the desire to control additional territory for economic, military or political purposes. Countries that have been defeated in wartime have often been obliged to cede control of territory to their victorious opponents. Following World War I, for example, the defeated Germans ceded East Prussia to Poland and the Alsace–Lorraine region to France.

Relationships between power and territory can be observed at all geographical scales. Power implies the opportunity to exercise control or influence over others, and control of territory is at times an objective of the exercise of power and at other times a result of its exercise. In the household, for example, a parent controls territory by placing gates or other barriers in order to deny toddlers access to hot stoves, fireplaces, and medicine cabinets. Older children, although not subject to territorial control through physical restraint, are admonished to remain within designated areas when not supervised by parents directly. For example, a ten-year-old may be told, "You may ride your bicycle on this block, but may not cross Main Street by yourself." Transgression of territorial limits on the part of the child brings swift punishment. Punishment reinforces the parent's power over the child – power exercised through territorial control, which, in most cases, is undertaken for the child's long-run benefit.

Exercise of power through territorial control continues throughout the life-cycle. In the workplace, certain rooms or areas may be kept locked or otherwise off limits to subordinate employees. "No Trespassing" signs limit the activities of hunters, fishermen and campers. Territorial control is a major prerogative of organized government. Local governments regulate land use through zoning laws, which restrict certain land uses to designated areas. Military bases are carefully guarded to prevent intrusion by outsiders, and tight security is maintained around embassies, consulates and other government offices in foreign countries.

In some societies, territorial control is exercised in order to privilege certain members of society above others. For example, South African apartheid policy denied members of that country's African majority the right to choose where they wished to reside. Not until the 1990s were Africans permitted to participate as full members of the South African political community. In the United States prior to the 1960s, African-Americans in the South were denied the right to attend integrated schools and otherwise associate with whites on an equal basis. Theaters, restaurants, sports stadiums and other public buildings were segregated, with African-American patrons restricted to undesirable, second-rate locations. In the United States, Australia and some other countries, indigenous peoples have been restricted to reservations, which usually consist of undesirable land of little value. In these and many other cases, explicit control of territory is linked with the use of power by one group over another.

2.4 Territory and conflict

Numerous wars throughout history have arisen as a result of disputes over control of territory. Why do states fight over control of territory? Certain territories are particularly desirable because of specific attributes or locational considerations. The presence of natural resources can make a territory especially desirable. For example, Iraq's takeover of Kuwait in August 1990 was occasioned in part by the presence of valuable petroleum reserves in Kuwait, while critics of the American response via Operation Desert Storm charged that America's military intervention was motivated largely by the desire to protect petroleum supplies.

Other territories are coveted for their strategic locations. Some are valuable for military purposes. Hence the Israelis have long desired control of the Golan Heights and the British have long maintained control of Gibraltar on the gateway to the Mediterranean. Other territories are valuable because their control can facilitate trade or economic growth. For example, Russia long desired an outlet to the Mediterranean. Hence the Russians have long undertaken efforts to obtain control of the Straits connecting the Mediterranean to the Black Sea on her southern border. Likewise, American control of the Panama Canal Zone enabled the United States to control shipping between the Atlantic and Pacific Oceans. Even when the United States ceded control of the Canal Zone to Panama in the late 1970s, the treaty stipulated that America could continue to dominate international trade in the region.

Throughout the course of history, relationships between power, territory and conflict have been affected by levels of military technology. Moreover, the effects of military activity on non-combatants has changed dramatically over time. In ancient and medieval times, military conflicts were settled by small armies. The large majority of the population were farmers whose day-to-day lives were seldom dramatically affected by armed conflict. The Biblical account of David and Goliath illustrates that ancient armies resolved conflict through individual combat, with few casualties. Today, in contrast, the twentieth century has seen far more deaths in battle than all previous centuries combined. Although the end of the Cold War has lessened the immediate threat of a nuclear holocaust, the technology to obliterate all of human civilization continues to spread.

2.5 Geopolitics and the modern world economy

Throughout history, authors from all over the world have identified and described the relationships between power, territory, conflict and location. Concepts relevant to geopolitical thought can be found in the writings of Aristotle, Confucius, Machiavelli and many other ancient and medieval authors. The formal study of geopolitics did not begin in earnest, however, until the late nineteenth century. Formal analysis of geopolitics coincided with the end of the Age of Exploration. By that time, the European powers had explored and begun to establish colonies in all of the inhabited portions of the world.

The modern world economy is characterized by capitalism, global economic interdependence and political fragmentation. The world economy as we understand it today began to emerge in Western Europe at the time of the Renaissance. As the world economy developed, the concept of a nation-state began to emerge. Nation-states were delineated territorially, and they linked cultures and ethnic groups to specific political units. By the end of the Middle Ages, many of today's major European nation-states, including England, France, Spain, Portugal, the Netherlands and the Scandinavian countries, had already come into existence.

At the same time that modern nation-states began to emerge, their economies became increasingly interdependent. The modern world economy, characterized by global economic interdependence along with a system of interdependent sovereign states, thus began to emerge in Europe during the Renaissance era. Since that time, the modern world economy has expanded to encompass the entirety of human civilization.

The modern world system stands in contrast to two previously recognized modes of international economic organization. These include mini-systems and world empires. A mini-system is a small, isolated society that lacks trade relationships with other societies. Typically, tribal groups in various pre-industrial societies such as the aborigines of Australia, pre-Columbian North and South American Indians and similar societies elsewhere in the world are considered mini-systems, although contemporary archeological evidence indicates that trade relationships among the "primitive" societies of pre-Columbian North America and traditional Africa were far more sophisticated and complex than would be typical of a mini-system in theory. For example, archeologists have discovered artifacts manufactured by the Mayans of Central America in sites in the present-day eastern and central United States; in other words, clear evidence that a sophisticated system of trade relationships existed among indigenous Native American cultures long before the arrival of European settlers. In recent years, scholars have questioned the concept of the mini-system. Even very ancient societies with very simple economies nevertheless maintained trade relationships with neighboring societies. Thus, trade relationships among less developed societies are better conceptualized in terms of a continuum of relationships, rather than comparing real societies to the abstract category of an isolated one (Chase-Dunn and Hall 1997).

World empires were political entities that achieved economic, military and political domination of large territories comprising many distinct nations. The Roman Empire, the Aztecs and Incas of the Americas, the Babylonians and Assyrians of the ancient Middle East, and the various dynasties of ancient and medieval China are examples of world empires. World empires emerged following the systematic conquest and subjugation of some nations by others. Conquerors forced subjugated cultures to contribute food, tax revenues, capital and labor to the central government as tribute.

As world empires expanded, they eventually became unstable. The collection of tribute over larger and larger areas required an increased bureaucracy as well as a larger and larger military establishment to enforce payment of tribute. As expansion continued, the costs of defending expanded borders became prohibitive, and eventually the world empire would collapse of its own weight.

The concept of a world empire is illustrated through examination of the Roman Empire, which began as a modest city-state in the vicinity of modern-day Rome. Over several centuries, Rome began to conquer neighboring city-states, and by 300 BC it had become dominant over much of modern-day peninsular Italy. Rome's victory over Carthage in the Punic Wars gave the now-expanding Roman Empire undisputed control over the central Mediterranean. By the time of Christ, the Romans had expanded throughout

Europe and the Middle East and into Africa and western Asia. As time continued, however, the Roman state began to weaken. For example, the Germanic tribes of northern and eastern Europe began to rebel against Roman domination. The empire began to decline – at first gradually and then more rapidly. Before AD 500, the Roman Empire had collapsed completely. The city of Rome, which had been the largest in the world during the height of the Roman Empire, declined precipitously as the Roman Empire collapsed.

2.6 The dawn of modern geopolitics

By 1900, the task of mapping and exploring the earth and its resources had largely been completed. All of the inhabitable or commercially valuable parts of the world had been divided into formal colonies controlled directly by the European powers (as in Africa) or into less formal spheres of influence, which were nevertheless subject to European economic and political control (as in East Asia). There were no additional new lands to explore and conquer, and the limits to potential European colonial expansion were now known.

Because the Age of Exploration had now passed into history, no longer could the European powers expand their resource bases through the incorporation of additional colonies outside Europe. Increasingly, the arena of conflict moved from outside Europe to Europe itself. It is no coincidence that the late nineteenth century – a time in which the earth's resources had been surveyed and the incorporation of the entire world into the European-dominated world economy had been completed – was the period in which formal geopolitical thinking emerged in Europe.

During the late nineteenth and early twentieth centuries, distinct schools of thought emerged in many different countries. In large measure, the geopolitical views of scholars in each country were closely intertwined with the foreign policy goals of that country. Common to each approach, however, was a concern with large-scale, systematic generalization along with particular emphasis on the role of the scholar's home country within the developing and ever-changing world political order. Here, we examine the geopolitics of Britain, France, Germany, Russia and the United States.

2.6.1 BRITISH GEOPOLITICS

In 1897, the United Kingdom celebrated the Diamond Jubilee – the 60th anniversary of the ascension of Queen Victoria to the throne of England. Queen Victoria remains a symbol of the heyday of nineteenth-century British imperialism. Victoria reigned over an empire that encompassed the globe. By the end of Victoria's reign, the sun truly never set on the British Empire. Britain had established herself as the dominant power within the world economy during the eighteenth century, and she maintained this domination during the nineteenth.

Establishment and maintenance of the vast and far-flung British Empire depended on British control of the seas. The location of Great Britain on an island off the mainland of Europe had long attuned the British to maritime activities, and the British Navy was far stronger than its European counterparts. British maritime power was seen to balance the larger populations and continental resources of Central Europe, especially Germany and Russia. In particular, the British were fearful of the growing military and economic power of Germany, whose power had expanded considerably following political unification in the mid-nineteenth century and as a result of the Industrial Revolution.

British concern with Continental domination of the world order was summarized by the well-known words of Sir Halford Mackinder, the leading geopolitical thinker in Britain during the early twentieth century. Mackinder summarized his view of geopolitics:

> Who rules East Europe commands the Heartland;
> who rules the Heartland commands the World Island;
> who commands the World Island commands the World.

By the Heartland, Mackinder meant the core of the Eurasian continent, including Germany, Eastern Europe and European Russia. Control of

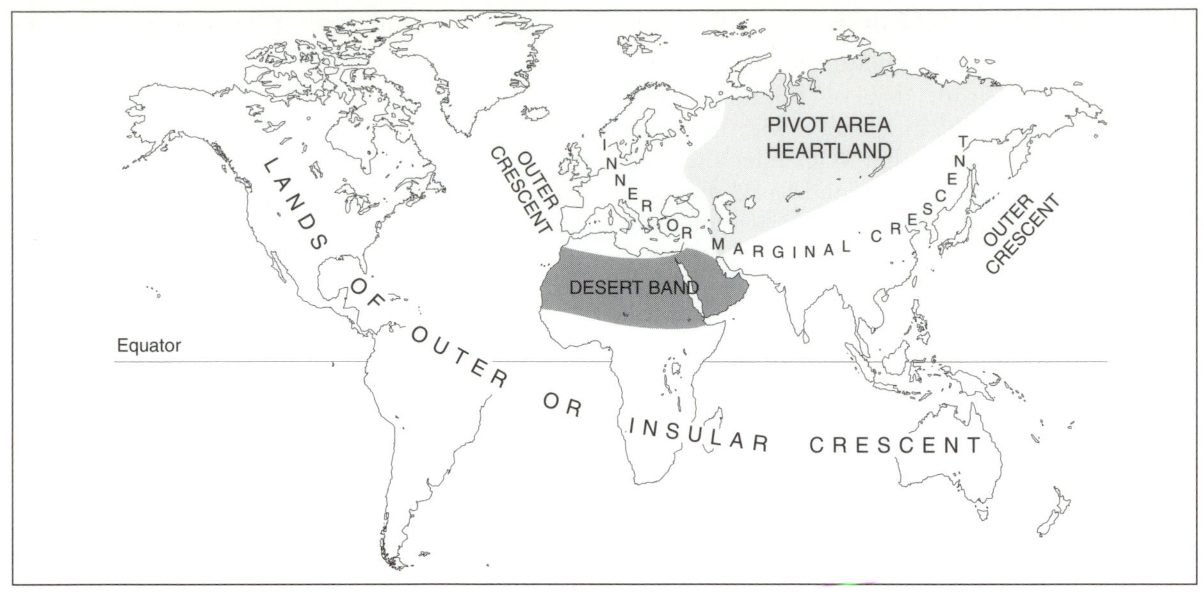

Figure 2.2 Mackinder's 'heartland modes' (source: Mackinder 1904)

the Heartland, in Mackinder's view, implied control of the World Island, or the great landmass of Europe, Asia and Africa (Figure 2.2). The geopolitical relationships among the Heartland, the World Island and the rest of the world can best be seen through examination of a map of the world drawn from a polar projection centered on Eastern Europe. Such a projection illustrates the extent to which the landmasses of the world are centered on Eastern Europe and Western Asia. This area was described by Mackinder as "the pivot of history".

How could the British balance the potential threat of Continental dominance in the World Island? Mackinder regarded world history in terms of recurring conflict between land-based and sea-based power. During the Age of Exploration, technological advances in shipping and naval activities along with European emphasis on colonialism and overseas expansion had tipped the balance of power in favor of the sea-based powers. By the nineteenth century, however, the Age of Exploration was coming to an end. The development of the railroad, the internal combustion engine and other technologies facilitating land-based transportation and communication were seen by Mackinder to shift the balance of power toward the land-based powers. The Heartland, secure from maritime attack but blessed with access to the heavily populated and resource-rich areas of China, India and the Middle East as well as Western Europe, was the natural center of land power.

Historically, the Russian Empire had been best situated to control the Heartland. By the end of the nineteenth century, however, Mackinder recognized that the growing might of Germany placed Germany, rather than the weaker Czarist state of Russia, at the center of the Heartland. On these grounds, Mackinder argued that it was incumbent on the British to dominate the world's oceans as a check on possible German expansion. Hence Mackinder argued that Britain should control the Rimland, or those areas of the world on and near the world's oceans. Allied victory in World War I, in which the sea-based powers of Britain and the United States vanquished the land-based power of Germany and her allies, seemed to bear out Mackinder's projections. Yet Mackinder argued presciently that Germany, despite her defeat in World War I, could again rise into a world power through control of the continental resources of the Heartland. As well, Mackinder stressed the importance of preventing a political or military alliance between Germany and Russia. Throughout the nineteenth century, the British took pains to retain Russia as an ally against Germany.

2.6.2 GEOPOLITICS IN FRANCE

The tradition of geopolitical thought in France differed somewhat from that of Britain. French geopolitics can best be understood with reference to the French position within Mackinder's scheme of recurring conflict between land-based and sea-based power. Germany and Russia represented the major land-based powers of the world, whereas Britain and the United States were the world's predominant maritime powers. France, on the other hand, was situated between the centers of land-based and sea-based power. Thus, France is the only European power that can be considered part of both the Heartland and the Rimland.

Despite the victory of the Allies in World War I, the French were concerned about the possible expansion of German power within Europe. For the most part, France supported those provisions of the Treaty of Versailles that imposed blame for the war on Germany, forced the Germans to relinquish her colonies and European territory to the Allies, and required the Germans to pay substantial reparations to the Allies. Resting uneasily behind the Maginot Line, many in France took pains to contrast the philosophies underlying French and German culture. The French strongly endorsed the League of Nations and advocated expanded international cooperation to settle disputes.

The French school of geopolitics took pains to establish contrasts between West and East. The Western tradition, epitomized by France as well as Britain and the United States, emphasized cooperation and flexibility. The East, symbolized by the feared Germans, represented authoritarianism and rigidity. Only in the extent of its colonial empire did the French surpass the Germans. The French thus regarded German views of territorial expansion with suspicion and alarm, while strongly supporting international efforts to negotiate international disputes through formal organizations such as the League of Nations.

2.6.3 GEOPOLITICS IN GERMANY

Both British and French geopolitics evolved in accordance with their respective countries' positions within the European world order of the late nineteenth and early twentieth centuries. Similarly, geopolitics in Germany can best be understood with references to German history and geography.

France and Britain, along with the other nations of Western Europe, had achieved political unity – in effect, having emerged as nation-states by the Renaissance. In contrast, the German-speaking areas of central Europe were characterized by deep political fragmentation until the middle of the nineteenth century. Present-day Germany consisted of hundreds of small states, principalities and petty kingdoms. Only under the dominance of Bismarck's Prussia in the mid-nineteenth century did Germany achieve political unification.

German unification proved to be a powerful springboard for the growth of German economic and political power. By 1900, Germany was the third leading industrial country in the world, behind Britain and the United States. Yet the geographical position of the united German state rendered Germany much more difficult to defend. Located at the center of the great European Plain, northern Germany had always been a crossroads, vulnerable to attack. Germany lacked the natural insularity of the British Isles, and was faced with traditionally hostile neighbors on both sides – France on the west, and the Eastern European powers along with Russia on the east. During the late nineteenth century, German foreign policy emphasized rapid territorial expansion in order to counter the possibility of attacks on both fronts. A strong, united Germany or Mitteleuropa including

Plate 2.2 Border between Germany and the Netherlands (photo: Kathleen Braden)

all of the German-speaking peoples of Central Europe would be the most effective means of preserving the integrity of German culture and precluding attacks on the German state by hostile neighbors.

The German defeat in World War I confirmed these views in the minds of many Germans. The Treaty of Versailles obliged Germany to recognize an independent Poland on its eastern frontier and to cede a substantial portion of East Prussia to the new Polish state. Poland, Czechoslovakia and other newly independent nations of Eastern Europe were established as a buffer between Germany and Russia. The establishment of the Polish Corridor separated East Prussia from the remainder of Germany. On its western flank, the Germans ceded Alsace–Lorraine, which they had won from the French in 1870, back to France. The territorial and military losses suffered by the Germans in World War I rendered the German state even weaker than it had been prior to unification. In response to this perception of vulnerability, German geopolitical thought re-emphasized the value of territorial expansion in conjunction with the unification of German-speaking peoples throughout Central Europe. Only through territorial expansion could the German state secure itself from external attack on both the western and the eastern fronts.

German geopolitics was influenced by the ideas of Friedrich Ratzel (1844–1904). Ratzel, who is often regarded as the founder of modern, systematic political geography, was influenced by Darwin's theory of evolution. Ratzel argued that states, like organisms, obey laws of evolution, with the principle of the survival of the fittest underlying competition between states. Strong states prosper and expand, while weak states decline and die. Thus Rudolf Kjellen (1846–1922), who invented the term *geopolitik*, described geopolitics as "the science which conceives of the state as a geographical organism or as a phenomenon in space".

The organic analogy underlies several of the key concepts of German geopolitics. Central to German geopolitical thought is the idea of *lebensraum* or living space. Expansion of territory under its control could help to ensure the long-run survival and competitive position of a state, and expansion was seen as a critical key to a state's growth and development. For post-World War I Germany, territorial expansion would result in a substantial increase in German power, blunting the threat of attack from Russia in the Heartland and from the oceanic powers of Britain and France to the west.

The fundamental principles of German geopolitics were expanded and refined following World War I by theorists at the Institut für Geopolitik in Munich, which was headed by Karl Haushofer (1869–1946). Influenced by the theories of Mackinder, Ratzel and Kjellen, Haushofer argued that a dynamic state required *lebensraum* to achieve autarky, or economic and political self-sufficiency. Autarky implies that a state requires adequate territory to ensure domestic access to raw material and markets in order to prosper and develop.

To an extent, autarky was equated with colonialism. The British, for example, had obtained valuable raw materials and trade advantages from its far-flung empire. German attitudes towards colonialism, in contrast, were somewhat more ambivalent, perhaps because Germany's colonial empire was neither as large or as productive as those of its European rivals. Moreover, the Treaty of Versailles stripped Germany of its colonies, which were ceded to the League of Nations and administered under League of Nations mandates awarded to Germany's enemies. The Germans recognized that any attempt to revive its colonial empire would meet with extensive resistance on the part of the other European powers. Hence it was argued that while the possession of colonies could indeed provide raw materials, markets for domestic products and outlets to accommodate displaced and surplus populations, the political and economic costs of maintaining a colonial empire were considerable. Thus Haushofer argued that it was in Germany's interests to develop a system in which she could extract the economic benefits of colonialism without absorbing its costs.

This goal could, in the opinion of Haushofer and other German theorists, best be achieved through territorial expansion into eastern and southeastern Europe. Kjellen, for example, argued that the expanded German state could include not

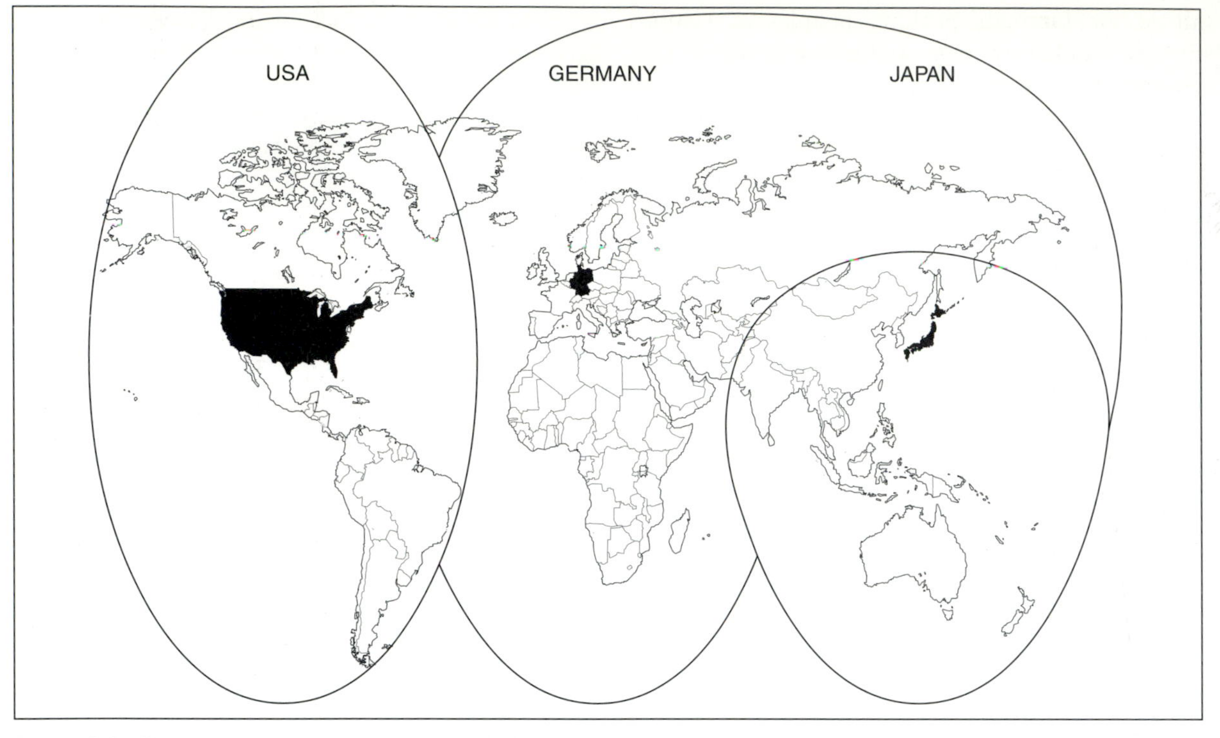

Figure 2.3 The world divided into three spheres of influence (after Nijmen 1992)

only eastern Europe from Poland to the Balkans, but expand onward into the Middle East and perhaps westward into Africa. This view of large-scale German expansion contributed to a third fundamental concept of German geopolitics – that of panideen or pan-regions. For Haushofer, a logical consequence of competitive expansion would be the development of a small number of pan-regions, each consisting of a large area of the world under the domination of one country. During the 1930s, a number of potential pan-regions were identified; for example, one conception of pan-regions divided the world into three major spheres of influence: Europe and Africa under the domination of Germany; Asia and Australasia under the domination of Japan; and the Americas under the domination of the United States (Figure 2.3). It is this conception of pan-regions that George Orwell satirized in his well-known novel, *Nineteen Eighty-Four*.

During and after World War II, Haushofer and his colleagues were condemned as providing intellectual inspiration to Hitler and the rulers of Germany during the war years. The extent to which Nazi leaders drew direct inspiration from Haushofer is uncertain, but Germany's aggressive territorial expansion in the 1930s led to the outbreak of World War II. In succession, Nazi Germany took over Austria and Czechoslovakia. The war was declared following the German invasion of Poland in September 1939. It is unlikely, however, that Hitler regarded Haushofer and the *geopolitik* school as a blueprint for foreign policy activities. Indeed, Haushofer pointed out that the German invasion of Russia in 1941 was inconsistent with geopolitical principles in that it forced the Nazis to conduct war on two fronts – a war which by 1945 they had lost. Nevertheless, the end of World War II brought about major changes in geopolitical thinking. Europe, once the center of the world economy and the chief locus of geopolitical discussion, was relegated to a dependent and indeed somewhat peripheral position in the world economy, while the United States and the Soviet Union emerged as the dominant economic and political powers in the world.

2.6.4 RUSSIAN GEOPOLITICS

Despite the demise of the Soviet Union, Russia remains the largest country in land area in the world. Russia's European territory, in fact, is nearly equal in size to the land area of the rest of that continent. The sheer size of European Russia has rendered it influential in international geopolitics for centuries, as Mackinder had recognized. Yet Russia's vast geopolitical potential was only beginning to be realized by the end of the nineteenth century.

Relative to the rest of Europe, the Russian state under the Czars was huge, peripheral, backward and isolated. Yet the vast and fertile Russian territory had long been targeted by foreigners for invasion. During the Middle Ages, Russia had been invaded by the German-speaking Teutonic Knights from the west and by the Mongols and other Asiatic nomads from the east. Meanwhile, Russia's economic development lagged far behind that of Western Europe; industrialization came late to Russia, which remained a feudal society long after Britain, France, Germany and the other European powers had become characterized by industrial capitalism.

Russian geopolitics are in large measure derivative of Russia's perception of herself as vulnerable, isolated and peripheral. This was recognized as early as the late seventeenth century by Peter the Great, who traveled extensively throughout Western Europe before ascending the Russian throne in 1689. After he became Czar, Peter took pains to establish Russia as a major European power by encouraging Western influence. He opened Russia to Western trade, and established the city of St Petersburg on the Baltic, making the new city capital of the Russian state. Peter and his successors established the traditional cornerstones of Russian international policy: secure borders, access to warm-water ports, elimination of economic dependency and expansion to the east.

Most of European Russia is a large and relatively featureless plain, which facilitates easy internal communication. To the north, Russia is bordered by the Arctic Ocean, while the great mountain ranges of Asia form its southern border. On its western flank, however, the Russian plain merges with the great plain of Central Europe, and this plain had been an important invasion route for centuries. Throughout Russian history, Russian leaders worked to counteract Germany's expansion tendencies by securing her western border. Soviet domination of Western Europe during the Cold War era can be regarded as a logical extension of this traditional policy.

Although Peter the Great was eager to promote trade between Russia and Western Europe, he recognized that Russian trading activities have always been hampered by climatic constraints and vast distances. The major rivers and ports of Russia are blockaded by ice for several months each winter. Only Murmansk in far northern Russia is an ice-free port, but its remote location on the Arctic Ocean renders it of little value for trade with Western Europe. Thus a major objective of Russian policy has been expansion to warm-water ports and trading opportunities. In particular, Russia desired control of the Black Sea and the Bosporus and Dardanelles Straits, connecting the Black Sea to the Mediterranean and consequently to the Atlantic. Russian control of this territory was an important issue during the Crimean War of 1853–1856. Throughout the eighteenth and nineteenth centuries, Russia expanded steadily to the southwest.

Paralleling this southern and western expansion has been an eastward movement through Siberia to the Pacific Ocean. Throughout the seventeenth, eighteenth and nineteenth centuries, Russian influence in Asia expanded steadily. The establishment of Vladivostok and other Pacific ports and the completion of the Trans-Siberian Railway helped to link Siberia with European Russia. Thinly populated Siberia stands in sharp contrast with the densely populated countries of eastern Asia to the south. Throughout the Middle Ages, Russia had experienced successive invasions of Asians, and the fear of continued Asian and particularly Chinese expansion has long been a dominant component of the Russian worldview.

The overthrow of the Czars and the subsequent establishment of the Soviet Union brought about some fundamental changes in Russia's approach to geopolitics. Marxist theory, which provided the impetus for the Russian Revolution, predicted that the excesses of capitalism would encourage a worldwide socialist revolution. The advanced industrial countries such as Germany and Britain were generally considered more likely candidates for socialist revolution than was backward, underdeveloped, agricultural Russia. After the Russian Revolution had established a communist government in the Soviet Union, Soviet leaders argued the merits of using Russia as a base from which to encourage worldwide socialist revolution as opposed to concentrating on the economic development of the Soviet Union itself.

This debate was critical to the struggle for power between Stalin and Trotsky, which ensued after Lenin, the first Soviet premier, died in 1924. Trotsky believed that the top priority of the Soviet Union was to facilitate an international socialist revolution; Stalin on the other hand committed himself to the policy of "socialism within one country". Stalin viewed the Soviet state as a socialist island surrounded by hostile, capitalist enemies. During the 1930s, Stalin sponsored large-scale industrial development programs and took drastic measures involving the deaths of thousands of political opponents to build up the economic and political power of the Soviet regime. The non-aggression pact between Stalin and Hitler, signed shortly before the outset of World War II, was seen to buy time for the Soviets to expand industrial production and build up military forces.

2.6.5 GEOPOLITICS IN THE UNITED STATES

Although geopolitics in the nineteenth and early twentieth centuries was focused primarily on Europe, the United States had emerged as one of the most powerful countries in the world by the time of World War I. In barely a century and a half, the United States grew from a European colonial outpost on the western shores of the North Atlantic Ocean to a leading military and economic power.

The United States enjoyed several substantial geographical advantages in its rise to international prominence. In contrast to the powers of Western Europe, America has enjoyed an abundance of natural resources, a large land area and secure borders. Its great distance from Europe had enabled America to remain neutral in most European conflicts. Throughout the nineteenth and early twentieth centuries, the United States expended a far smaller percentage of its resources on its armed forces relative to the nations of Europe. This in turn freed a greater proportion of governmental resources for civilian purposes, expediting the pace of American industrialization and economic development.

Throughout American history, United States foreign policy has shifted between introverted cycles, in which American interest in foreign policy issues has been sublimated to domestic policy concerns, and extroverted cycles when the United States took a more active interest in international relations. Thus introvert phases represent periods dominated by a philosophy of American isolationism, while extrovert periods are characterized by an attitude of intervention in foreign affairs.

Between the granting of American independence following the Treaty of Paris in 1783 and the end of World War II, three extrovert phases and three introvert phases are recognized. The extrovert phases include the early years of American independence prior to 1825, the period between 1845 and 1867 when America completed its territorial expansion across North America, and the period from the late 1890s until the end of World War I, when the United States first emerged as a global power. The period during and after World War II represents a fourth extrovert phase. The intervening periods are recognized as introvert phases.

After the Revolutionary War, America's main foreign policy concerns included ensuring sovereignty over her territory and removing European influence from the New World. Achievement of these objectives brought the newly independent United States into periodic conflict with Britain and France, including the War of 1812. The Louisiana Purchase of 1803 helped to secure American sovereignty over eastern and central North America, while the purchase of Florida from Spain in 1819 eliminated Spanish colonization of the present-day United States. In 1823, the

Monroe Doctrine was announced. The intent of the Monroe Doctrine was to establish as a cornerstone of American foreign policy opposition to any further European colonial expansion in North and South America. Geopolitical domination of the Americas has remained a central tenet of American geopolitics ever since.

During the second extroverted phase, the United States expanded across the North American continent. In the 1840s, America annexed Texas, acquired the Southwest following victory in the Mexican War, and obtained title to the Pacific Northwest by treaty with Britain, completing American extension of sovereignty to the Pacific Ocean. America's continental expansion was completed with the Gadsden Purchase of southern New Mexico and southern Arizona in 1853 and the purchase of Alaska from Russia in 1867.

The latter half of the nineteenth century was devoted primarily to the expansion of settlement across these new territories and the development of American industry. At the end of this extended introvert phase, the United States had emerged as the world's leading industrial power. At the end of the nineteenth century, however, the United States entered its third extrovert phase. For the first time, America became a colonial power. The previously independent kingdom of Hawaii was annexed, and the territories of Cuba, Puerto Rico and the Phillipines were acquired from Spain following America's victory in the Spanish–American War. At the same time, America took a more active role in establishing and maintaining economic and political dominance of Latin America. In the early twentieth century, American troops or warships were at various times deployed in Cuba, Mexico, Panama, Honduras, the Dominican Republic, and several other Central and South American nations, and the Monroe Doctrine was invoked to justify these incursions.

The extrovert phase beginning in the 1890s culminated in American entry into World War I. America initially remained neutral in the European conflict, and President Woodrow Wilson used the slogan "He kept us out of war" to secure re-election in 1916. However, Wilson's concern for a continued balance of power in Europe tilted American foreign policy toward Britain, resulting eventually in American declaration of war on Germany in 1917. American entry into the war secured an Allied victory. After armed hostilities ended, Wilson's vision of a "just peace" overseen by the League of Nations was incorporated, at least in principle, into the Treaty of Versailles in 1919.

Following the war, however, America renounced the internationalist, extrovert stance that it had developed over the two previous decades. Instead, the United States swiftly moved into an introvert phase in its foreign relations. The United States Senate refused to ratify the League of Nations Covenant by the two-thirds majority required by the Constitution. Although America had never previously restricted immigration, restrictive immigration laws were enacted in the early 1920s and high tariffs restricted American exports.

As international tensions mounted in Europe during the 1930s, many in the United States came to question the isolationist perspective that had dominated American policy since the end of World War I. Americans became increasingly divided over the relative merits of isolation and intervention. In general, intervention was favored in the Northeast and the South, while residents of the Middle West and the West leaned toward isolationism (Trubowitz 1998). In 1940, President Franklin D. Roosevelt promoted the establishment of a military draft – the first peacetime conscription in American history – along with Lend–Lease assistance to British and Soviet military interests. Both proposals were highly controversial, and both passed Congress only by narrow margins (Barone 1990).

On December 7, 1941, the Japanese attack on Pearl Harbor in Hawaii prompted an American declaration of war against the Axis Powers (Japan, Germany and Italy). Before the cessation of hostilities in 1945, the United States fielded over 15 million troops and suffered nearly one-third of a million battle deaths. The destruction of Hiroshima and Nagasaki by atomic bombs demonstrated the immense power of nuclear weapons. The end of the war in 1945 left America the world's strongest military and economic power. However, America did not retreat into isolationism as it had done following World War I. The

United Nations Charter was ratified by the Senate by a vote of 89 to 2 in September of 1945.

During the late 1940s the United States expanded economic and military aid to Western Europe through the establishment of the Marshall Plan and the North Atlantic Treaty Organization (NATO). By the end of the decade, a bipolar view of international relations, contrasting Soviet communism with American democracy, was characteristic of American foreign policy. The advent of the nuclear age and an increasing arms race between the United States and the Soviet Union ushered in a new era of geopolitics. The fourth extrovert phase lasted from the end of World War II until the late 1960s, when mounting American opposition to the war in Vietnam generated a new introverted phase. In the next chapter, we shall discuss American geopolitics within the context of the Cold War in greater detail.

The shift between extrovert and introvert phases in foreign policy mirrors ongoing tension between proponents of isolationism and internationalism – a debate which has lasted since America first achieved political independence two centuries ago. Generally speaking, isolationist views suggest that America should avoid interference in international affairs, except to the extent that the activities of foreign powers impact the United States directly. The internationalist approach argues that more direct involvement in international politics is the best way to secure peace in America. In general, isolationist views have tended to be dominant during introvert phases, while internationalist views have been more dominant during extrovert phases.

In addition to the tension between isolationism and internationalism and the Monroe Doctrine, two other considerations have influenced twentieth-century American geopolitics – the role of aviation and the role of the Arctic regions. Throughout the twentieth century, American foreign policy has emphasized American dominance of the skies and, more recently, outer space. The importance of the airplane to American geopolitics can be observed by constructing a polar map projection centered on the Northern Hemisphere (Figure 2.4). A polar projection illustrates the fact that the shortest air distance between two places is the "great circle" route, which cuts across parallels of latitude. The shortest distance between London and New York, for example, is a great circle across Newfoundland and Greenland. Between London and Los Angeles, or between Moscow and New York, the great circle cuts directly across the Arctic.

Corollary to American emphasis on the importance of the airplane in international politics is emphasis on the importance of the Arctic regions, which are located along the shortest air routes between Eurasia and North America. Vilhjalmur Stefansson, for example, regarded the Arctic as an "American Mediterranean": as he put it, "The so-called Arctic Ocean has the nature of a mediterranean sea: the land masses of Europe, Asia and North America are grouped around it somewhat as Africa, Europe and Asia surround the sea that bears the name Mediterranean." The term mediterranean, of course, means "middle of the earth", and thus Stefansson conceptualized the Arctic Ocean as being in the middle of the United States's sphere of influence. Only the small and friendly country of Canada stood between the United States and domination of the High Arctic.

Well known for his exploration of polar regions and his extensive stays in the High Arctic, Stefansson was convinced that the region could support a large population and extensive economic development. Other American writers, although doubtful about long-run economic and settlement prospects in the Arctic, nevertheless agreed that the advent of the airplane had substantially increased the importance of the Arctic region. During the Cold War, the Arctic was considered very important to American defense. Numerous military bases and missile tracking stations were operated in Alaska, northern Canada, Greenland and Iceland.

2.7 Conclusions

In this chapter, we have identified geopolitics as an important subset of political geography. The formal study of geopolitics began in the late nineteenth century – a time in which the process of colonialism by which the non-European world

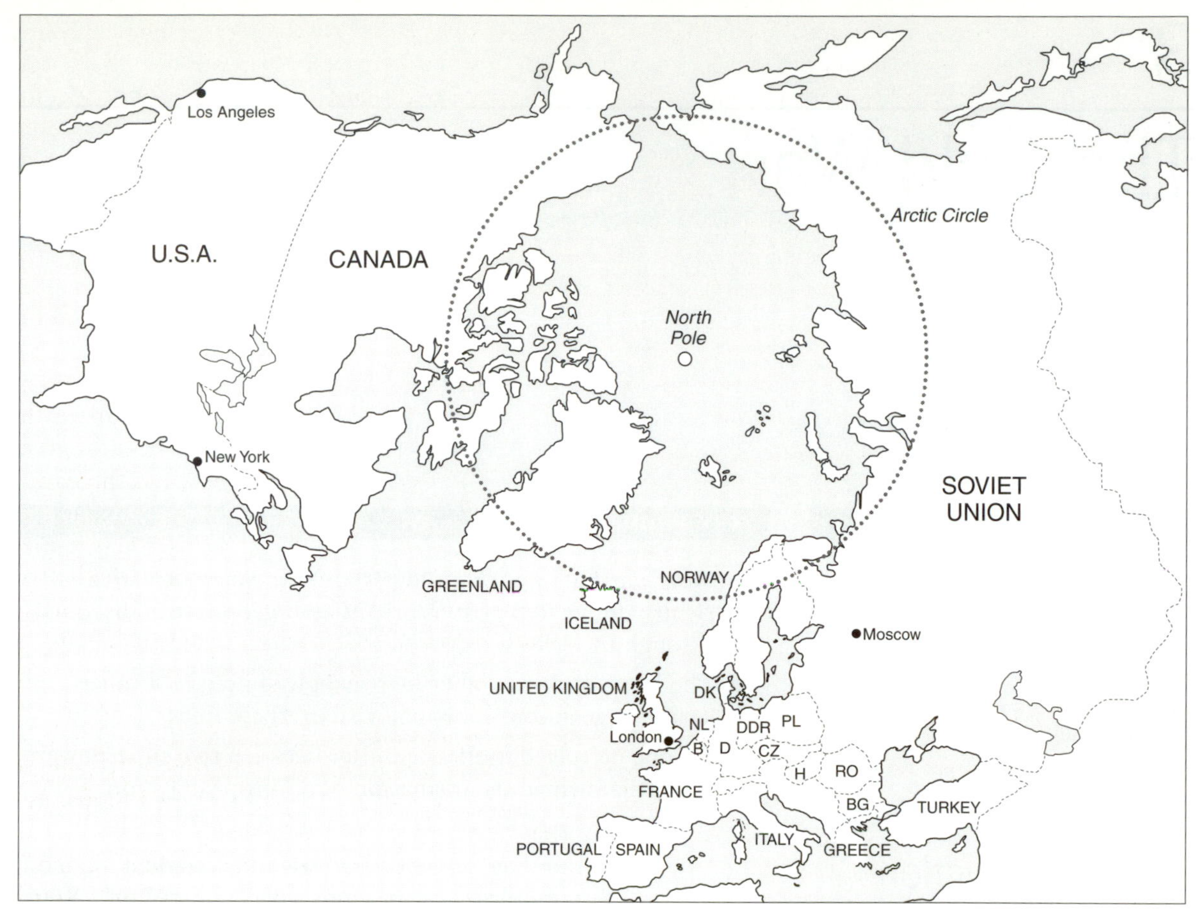

Figure 2.4 Polar projection of the world (from K Dodds, *Geopolitics in a Changing World*, (Prentice Hall, 1999))

became integrated into the European-centered world economy had reached its limits. British, French, German, Russian and American geopolitical thinkers developed theories that in each case informed the foreign policies of their respective countries.

Before the twentieth century was half over, however, Europe had been devastated by the two greatest wars in history. No longer was Europe at the center of the world's geopolitical stage. Instead, global geopolitics were dominated by two powers outside Europe: the United States and the Soviet Union. For more than 40 years, global geopolitics were dominated by conflict between these two superpowers. The Cold War – accompanied by the development of technology that made it possible for human beings to obliterate their own species with the push of a button – had profound consequences for the geopolitics of Europe and throughout the world.

3

The Cold War

KEYWORDS

blocs,
domino theory,
containment,
détente,
glasnost,
shatterbelt,
non-alignment

KEY PROPOSITIONS

- New war-fighting technologies and the Soviet/American rivalry dominated post-World War II geopolitical discourse through the 1980s.
- The locus of this conflict increasingly shifted to a third world stage and to four major shatterbelt areas.
- Europe perceived itself as a buffer between two superpowers, a fact that helped the formation of a European Union.

3.1 Introduction

The end of World War II in 1945 brought about substantial and profound changes in international relations. Since the Renaissance, the world economy had been centered in Europe, and geopolitical theory was concerned primarily with European states and their interrelationships. The conclusion of World War II signified that Europe was no longer at the center of the global political economy. After the war, the United States and the Union of Soviet Socialist Republics emerged as the world's leading powers, while the countries of Europe were dependent on American and Soviet aid to rebuild their war-torn and devastated economies.

Conflict between the Americans and the Soviets soon dominated geopolitical thought. Within a very few years, geopolitical and ideological differences between America and Russia were to divide Europe and initiate the Cold War. Sir Winston Churchill's famous phrase, "an iron curtain has descended across the continent" came to symbolize Europe's position in post-World War II geopolitics. Europe was no longer in a position to initiate geopolitical change; rather it was a potential battleground, both militarily and economically, between the competing interests of the United States and the Soviet Union. By the 1950s, Western Europe had become incorporated into the American-dominated North Atlantic Treaty Organization (NATO), and thousands of American troops were stationed in Germany and elsewhere in Europe with the intention of protecting Western Europe from potential Russian attacks. Similarly, the Warsaw Pact symbolized the incorporation of Eastern Europe into the Soviet sphere of influence.

The world's geopolitics were also altered radically by new military technologies, which drastically reduced the distance between competing nations. The American detonation of the atomic bomb at Hiroshima and Nagasaki in 1945 for the first time unveiled the destructive power of atomic weaponry. By the early 1950s the Russians had also developed and tested nuclear weapons, and the wholesale destruction of civilization through atomic warfare became possible for the first time.

3.2 The emergence of the Cold War

During World War II, the common goal of defeating Nazi Germany had placed the Soviet Union in an uneasy alliance with the United States, Britain and their Western allies. Allied military success left the United States, Britain and the Soviet Union as the world's three strongest countries. Within a short time after the conclusion of the war, however, the USSR and the United States had become bitter enemies, and the Cold War was in full swing.

Why the United States and the Soviet Union moved so quickly from allies to antagonists represents an interesting historical question. By the beginning of 1945, several months before actual fighting ceased, it was clear that the Allies would win the war. The three dominant members of the Allied coalition, of course, were the United States, the Soviet Union and Britain. Wartime strategies had been developed jointly by the leaders of these countries: Franklin Delano Roosevelt, Joseph Stalin and Winston Churchill. As Allied military success became assured, new geopolitical relationships among the now dominant Americans, Soviets and British arose. Why then, did the new world order arising after 1945 come to be dominated by competition between the Americans and the Soviets, with the British, in effect, serving as a junior partner of the United States?

In order to address this question, we must consider the fact that the Cold War was one of five possible geopolitical scenarios. This leaves us with four alternatives. One alternative would have been mutual cooperation among all three powers. This alternative would imply continued cooperation among the Allied leaders and a strong international role for the United Nations. The second alternative would have been continued mutual antagonism among the wartime allies. Carried to its logical conclusion, this alternative could have generated three pan-regions reminiscent of those proposed by German geopolitical thinkers prior to the war, with the Old World pan-regions dominated by Britain and Russia rather than by Germany and Japan. The three remaining possibilities involved alliances between two of the three powers. A Soviet–American alliance against British imperialism and a Soviet–British alliance against American economic power were plausible alternatives to what eventually developed – the Anglo-American alliance against communism and the resulting Cold War.

Recent scholarship has investigated the role of Britain in the creation of the Anglo-American alliance (Taylor 1990). Although Churchill's Conservative Party was defeated by its Labour opposition shortly after VE Day, the new Labour government did little to alter the basic course of British foreign policy. Massive American aid proved valuable to British attempts to rebuild that country's war-shattered economy. Recognizing that the continued flow of American aid was dependent on the cementing of a military alliance between America and Britain, a majority within the British Parliament supported an Anglo-American alliance. Within Britain, both major parties supported Parliamentary legislation accepting American loans. Although the proposal was vigorously opposed by a coalition of left-wing Labour and right-wing Conservative backbenchers, the legislation passed the House of Commons by a vote of 345 to 98. Thus "a vital link was slotted into place in the US hegemonic edifice" (Taylor 1990: 85). Those supporting the approval of an Anglo-American alliance were motivated primarily by the expectation that American financial assistance would prove critical to the return of British prosperity.

The United States, meanwhile, moved into an extrovert period in its foreign policy. Most in the United States supported the United Nations and became increasingly critical of Soviet communism, especially after Eastern Europe and China came under communist control in the late 1940s. By the early 1950s, a broad-based, bipartisan consensus on the desirability of a strongly internationalist Cold War foreign policy had emerged in the United States. This consensus dominated the politics of American foreign policy until it unraveled as a result of controversy over American involvement in Southeast Asia in the late 1960s.

3.3 The unfolding of the Cold War

By 1950, escalating tensions between East and West had generated the long hostility known as the Cold War. As the Cold War deepened, the states of Europe came to be divided into two camps: the Eastern European bloc including Poland, Czechoslovakia, Hungary, Yugoslavia, Romania and Bulgaria under the domination of the Soviet Union; and a Western bloc including Britain, France, Spain and the Low Countries under the leadership of the United States. The eventual division of Germany into communist-dominated East Germany and Western-oriented West Germany perhaps best symbolized the increasing bipolarity of Europe in the 1950s, as did the division of the former imperial capital of Berlin into Soviet and Western-dominated spheres of influence.

By the late 1940s, many American leaders had become concerned that the Soviets would continue to extend their influence beyond Eastern Europe. The "domino theory", which suggested that countries were successively vulnerable to communist influence like a row of falling dominoes, was often proposed as a metaphor for possible communist takeover of additional countries. In response, President Harry S. Truman established the Truman Doctrine, in which the United States pledged military and economic assistance to European countries threatened by communist takeover. In effect, the Truman Doctrine served to legitimize American interests in European politics. These interests were reinforced by the Marshall Plan, in which millions of dollars in American foreign aid were sent across the Atlantic to help war-torn European countries rebuild their economies. Soon after the Truman Doctrine was established in 1947, aid was provided to fight communist insurgency in Greece and Turkey. In Germany, the United States supported an airlift to counteract an Eastern blockade of West Berlin. NATO was established to coordinate military activities among the United States and its Western European allies.

By no means was the increasing hostility and competition between the United States and the Soviet Union confined to Europe. The American-supported Nationalist government of China fell to Mao Zedong and the Communists in 1949, and Nationalist leader Chiang Kai-Shek fled to Taiwan. In 1950, Soviet-supported North Korean forces invaded South Korea. Led by the United States, the United Nations dispatched thousands of troops to Korea. The Korean War lasted three years and resulted in the deaths of 53,000 American soldiers.

3.4 The Cold War in the 1950s and 1960s

By the middle of the 1950s, the United States and the Soviet Union had become locked in an increasing arms race. In 1957, the Soviet Union announced the launching of its first orbiting satellite, Sputnik. The launch of Sputnik sparked Western fears that the Soviets were prepared to outdistance the United States in technological development. "Gaps" in American military technology relative to the purported capability of the Soviets were reported in the press, and the so-called "missile gap" became an important issue in American election campaigns during the late 1950s and early 1960s. The launch of Sputnik not only raised fears that Soviet technology was outstripping American performance, but in the minds of many Americans it represented a direct challenge to American domination of the air, which as we have seen represented a fundamental component of the American geopolitical worldview. Fears of falling behind the Soviets in military technology led to a substantial effort to improve American technological performance. The American space program was strengthened – leading eventually to the American moon landing in 1969 – and public education in mathematics, science and technology was buttressed by large flows of Federal revenues.

As the Cold War deepened, relationships between the United States and the Soviet Union continued to deteriorate. Each side used increasingly sophisticated technologies to monitor the activities of the other. Both established military bases in the territories of their allies. The United States, for example, set up military bases in Turkey and used them to gather intelligence about Russian military activities. In 1960, an American U-2 reconnaissance plane was shot down over Soviet air space. The Soviets accused the American military of spying and canceled an imminent summit meeting between Soviet Premier Nikita Khrushchev and the United States President Dwight D. Eisenhower.

In the early 1960s, a series of confrontations brought the United States to the brink of armed conflict with the Soviet Union. In 1961, Khrushchev demanded that Western troops be withdrawn from West Berlin. In August of that year, the East German government constructed a wall across Berlin to prevent East German citizens from fleeing to the western sector. For a while, American and Soviet tanks faced each other across the wall. The tension was broken only when the West agreed to let its personnel be inspected before they entered Berlin. The Berlin Wall remained in place until 1989, a symbol of the Cold War.

In 1962, another crisis developed over Cuba: in 1959, the Cuban government had been taken over by Fidel Castro. Castro turned Cuba into a socialist state and established close ties with Khrushchev and the Soviet Union. Castro's hostility toward the United States deepened when the Americans sponsored an unsuccessful attempt to overthrow his government via the Bay of Pigs invasion in 1961. By the summer of 1962, the Soviets had begun to send missiles to Cuba – less than 150 miles (240 km) from Miami, Florida. Air photographs taken in October 1962 showed the beginnings of the Soviet military installations, despite Soviet claims to the contrary. Rejecting the notion of a mutual withdrawal of missiles from Cuba and Turkey, the United States issued an ultimatum to the USSR about removing the Cuban missiles. On October 22, President John F. Kennedy gave a televised speech in which he announced that the United States would establish a naval blockade of Cuba. Ships headed for Cuba were to be searched, with military equipment quarantined. The crisis receded when Kennedy told Khrushchev that the United States would not invade Cuba if the missiles were withdrawn, and the Soviets agreed. Eventually, American missiles in Turkey were also withdrawn.

Although the effect of the Cuban missile crisis on the deterrence policy of the superpowers is not known, the years following 1962 were characterized by a significant lessening of tension between the Americans and the Soviets. In 1963, the two countries signed a partial test ban treaty, which outlawed atmospheric testing of nuclear weapons. Additional treaties were signed in 1966 and 1968. These banned the introduction of nuclear weapons into outer space and limited the spread of nuclear technology to non-nuclear nations.

While direct conflict between the United States and the Soviet Union certainly was reduced in the 1960s, the locus of superpower conflict shifted increasingly to the Third World. During the late 1950s and early 1960s, the European colonies of Africa, Asia and elsewhere were granted political independence. Many became battlegrounds of the Cold War, with East and West backing rival factions in internal power struggles. The Americans, Soviets and Chinese – who by the early 1960s had broken relations with the Soviet Union and were anxious to establish themselves in a leadership position within the Third World – donated large amounts of military and economic assistance to these newly independent nations.

3.5 Vietnam and the end of the Cold War consensus

During the 1960s, America became involved in the Vietnam conflict. Vietnam was split between the communist-dominated North Vietnam and a Western-oriented regime in South Vietnam. In 1964, the United States Senate adopted the Gulf of Tonkin Resolution authorizing Presidential military authority in South Vietnam. Only two senators opposed the Resolution, which was used to justify the commitment of nearly 400,000 American troops in Vietnam by the end of 1966.

As American military activities in Southeast Asia expanded, however, President Lyndon B.

Plate 3.1 Federal building in Washington, DC, with the American flag (photo: Kathleen Braden)

Johnson's decision to commit large numbers of troops and military supplies to Vietnam came to be questioned increasingly by the American public. By the late 1960s, a majority of Americans opposed further involvement in Vietnam. Over 250,000 Americans demonstrated against the Vietnam War in Washington, DC, in November of 1969, with thousands more participating in anti-war demonstrations elsewhere. As the 1970s began, it became clear that American military victory in Vietnam could not be achieved, and the Americans gradually turned over conduct of military operations to the South Vietnamese army. The Vietnam war ended with the fall of Saigon in 1975. By that time, American activities in Vietnam had resulted in nearly 60,000 American military casualties.

The heated and acrimonious debate over Vietnam affected American politics substantially. The Vietnam debate generated a new geographic split in American foreign policy. The Northeast, the upper Middle West and the West tended to be "dovish" regions, opposing American involvement in Vietnam; while the southern half of the country, which contained a large majority of military bases and personnel, was more supportive. Thus the Vietnam debate presaged a new Rustbelt/Sunbelt cleavage in American foreign policy (Trubowitz 1998), while it signaled an end to the extroverted period in American foreign policy that had begun with the bombing of Pearl Harbor in 1941. Instead, the United States entered an introvert cycle, which remained until the end of the Cold War. Opposition to foreign involvement was especially prevalent in the northern half of the country. The Sunbelt, on the other hand, has most consistently supported intervention – in part because the economy of this region benefited so substantially from the high levels of military expenditure committed by the Federal government during the Cold War.

3.6 The Cold War during the 1970s and 1980s

As the Vietnam build-up drew to a close, the Americans pursued a policy of *détente*, or released tension in US–Soviet relations. The idea of *détente* was to promote increased trade, cultural exchanges, openness in international relations and arms control. In the early 1970s, the Americans and Soviets initiated negotiations to limit the build-up of their nuclear arsenals. The Strategic Arms Limitation Talks (SALT) began in 1970, with the SALT I agreement signed by President Richard Nixon and Soviet Premier Leonid Brezhnev in October 1972. There were two basic sections to the SALT accord: an ABM Treaty, which specified that the two superpowers would limit their anti-ballistic missile (ABM) systems to one area and

Plate 3.2 Leningrad (now St Petersburg) and 1970s Cold War urban decorations to celebrate the Great October Revolution holiday: Leonid Brezhnev, then chair of the Communist Party of the Soviet Union (photo: Kathleen Braden)

Plate 3.3 As in Plate 3.2, but showing V. I. Lenin, founder of the Bolshevik Revolution (photo: Kathleen Braden)

eliminate further testing and development of ABM systems; and an agreement placing limits on each country's arsenal of anti-ballistic missiles.

Despite the signing of SALT I, the arms race continued to build throughout the 1970s. Both superpowers experienced unprecedented peacetime military build-ups in the late 1970s. American response to Soviet activity was fueled by the Team B report of Soviet military activities sponsored by George Bush, then Director of the Central Intelligence Agency. The Team B report indicated that the United States had significantly underestimated the extent of the Soviet military build-up. The Soviet invasion of Afghanistan in 1979 was regarded as evidence of Russian willingness to seek continued military superiority over the West.

In 1981, Ronald Reagan was elected President. His campaign platform was critical of his predecessor, Jimmy Carter, on the grounds that the Carter Administration had mismanaged America's defenses and had underestimated the growing military strength of the Soviets. Reagan took office with a determination to increase military spending. Spending on nuclear weapons doubled from $35 billion in 1981 to $70 billion in 1987, and total military spending peaked at $295 million in 1985 (see Chapter 6).

In the mid-1980s, surprising internal changes in the Soviet Union had the effect of reducing tensions with the West. Brezhnev died in 1982, and his immediate successors, Yuri Andropov and Konstantin Chernenko, also died shortly after assuming leadership of the Communist Party. In 1985, Mikhail Gorbachev was selected as General Secretary. At the age of 54, Gorbachev was the youngest man to hold this position since Stalin's time.

Within three years, Gorbachev had embarked on a sweeping and unprecedented series of domestic and foreign policy reforms. He promoted a policy of glasnost, or openness, by easing restrictions on emigration, censorship and dissent. Extensive economic reforms, encouraging limited development of private initiatives, were undertaken. In 1989, non-members of the Communist Party were permitted to run for public office against Party-sponsored candidates, and many were elected. Gorbachev sponsored the withdrawal of Soviet troops from Afghanistan, and signed an arms control agreement concerning European missiles with the West. During the summer of 1989, Gorbachev and United States President George Bush announced plans to reduce missiles and other armaments in Europe. International tensions lessened, yet few in the late 1980s were aware that the Cold War was about to come to a sudden and abrupt end.

3.7 Europe during the Cold War

As we have seen, the world geopolitical order between the late 1940s and the late 1980s was dominated by Cold War competition between West and East – the United States, Western Europe and their allies versus the Soviet Union, Eastern Europe and their allies. How did the Cold War affect other portions of the world?

As our previous discussion of post-1945 Britain has illustrated, World War II resulted in a shattering of the European economy. The war itself had resulted in millions of battlefield casualties and wholesale destruction of many cities and towns. Six years of warfare had drained the economic resources of the European countries. Moreover, Europe lost its dominant position in the world economy and in global geopolitics. The United States and the Soviet Union had emerged as the world's dominant powers. This, coupled with escalating nuclear and conventional arms development, induced many Europeans to perceive Europe as a battleground between the competing political and economic interests of the Soviets to the east and the Americans to the west. The Iron Curtain, along with evidence of the development of increasingly sophisticated and powerful nuclear weapons, lent credence to this belief in the minds of many Europeans.

By the late 1940s, the European economies had begun the long, slow process of economic redevelopment. American aid provided through the Marshall Plan encouraged the redevelopment of European industry. Yet Europe remained well behind the United States in productivity, economic development and global power. By the 1950s, some Europeans had begun to argue that economic and possible political unification would improve Europe's competitive position within the Cold War world economy.

Thus the stage was set for the unification of Europe. The first formal step in the European unification process was the establishment of the European Coal and Steel Community in 1951. This international agreement eliminated trade barriers in these vital industries. In 1957, six Western European countries – France, West Germany, Italy, the Netherlands, Belgium and Luxembourg – expanded the European Coal and Steel Community to form the European Economic Community or "Common Market". The intent of the European Economic Community was to eliminate internal trade barriers and establish common tariff policies.

The European Community expanded from its original membership during the ensuing decades. In 1973, the United Kingdom, Ireland and Denmark joined the Community, formally known as the European Community (EC). Greece joined the European Community in 1979, and Spain and Portugal did so in 1986. In 1995, Sweden, Austria and Finland joined the Community, which is now known as the European Union (EU). Today, the population, economic power and income of the Community is equivalent to that of the United States, and well exceeds that of the former Soviet Union. While expanding in size and economic power, the EU also expanded its role in Europe from strictly economic matters to many others. By the late 1970s, plans for a European Parliament were in place, with the first elections for seats in the European Parliament held in 1984.

During the 1980s, the pace of European unification accelerated. In 1985, the EU established a program to complete the development of a single internal European market by 1993. In 1991, the member states signed the Maastricht Treaty. The Maastricht Treaty included several provisions associated with increased political and economic unification. These included the establishment of a common police force, stronger military cooperation, a common system of telecommunications, and a central banking and currency system by 1999. After the Treaty was signed, however, it became evident that support for the Maastricht Treaty provisions was at best lukewarm in many parts of Western Europe. Opposition to further unification was evident when the treaty was subject to ratification within the states of the European Union. The voters of Denmark rejected it, while those in several other countries, including France, approved it by very narrow margins.

Why has opposition to the economic and political unification of Europe increased in recent years? Several reasons may be advanced. Some feared that the European Union might come under domination by Germany in a manner reminiscent of past German visions of a European pan-region. Ironically, the end of the Cold War has had the result of intensifying these fears. Prior to the end of the Cold War, the largest countries of the European Union (Great Britain, France, West Germany and Italy) were approximately equal in land area, population, and economic power. Once East Germany and West Germany reunified, however, Germany was clearly larger and more powerful than the others.

Other opponents of increased European unification cited alternative rationales. For example, environmentalists in Denmark attempted to restrict the import of German beer that was not bottled in recyclable containers, but were not permitted to do so under European Union free-trade rules. In particular, member states ran into difficulty in establishing a common currency and monetary unit. Naturally, each of the stronger member states wanted its currency to become the basis of the European monetary system. Historical distrust was

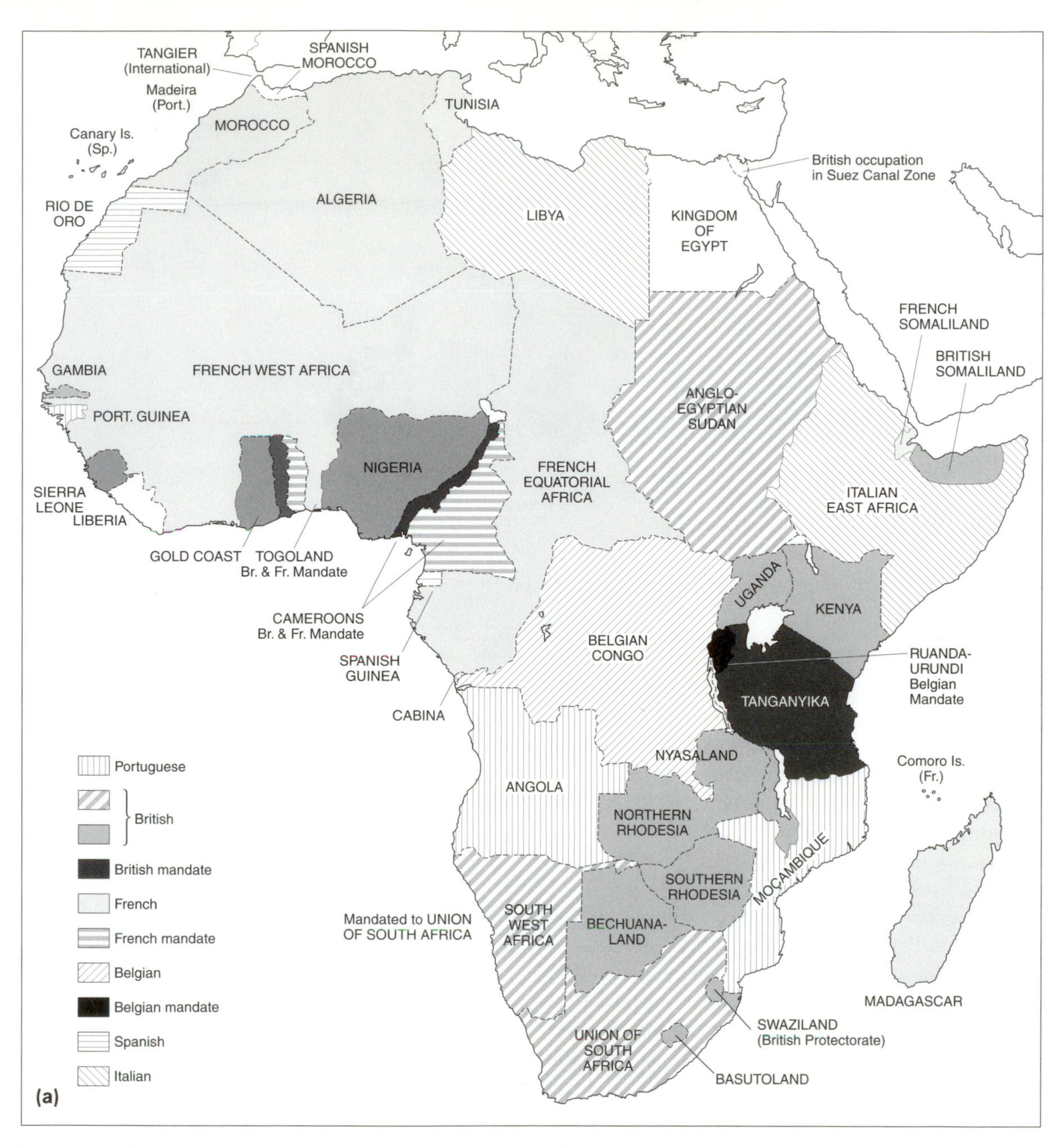

Figure 3.1 African and Asian colonies in 1939 (a, b); Chronology of independence (c, d) (from M Chamberlain, The Longman Companion to European Decolonisation in the Twentieth Century (Addison Wesley Longman, 1998))

fueled by the fact that those countries whose own currencies were strong relative to the European currency at the time of currency unification would benefit at the expense of those whose currencies were weak. Other critics pointed out that unification would undercut the opportunity for members to promote individual foreign policy goals.

The increasing prominence of nationalist movements in Europe may also have contributed to European skepticism about further unification. The internationalist commitment in European policy is tempered by strong nationalist movements in many countries. Demands for increased regional autonomy and in some cases partial or

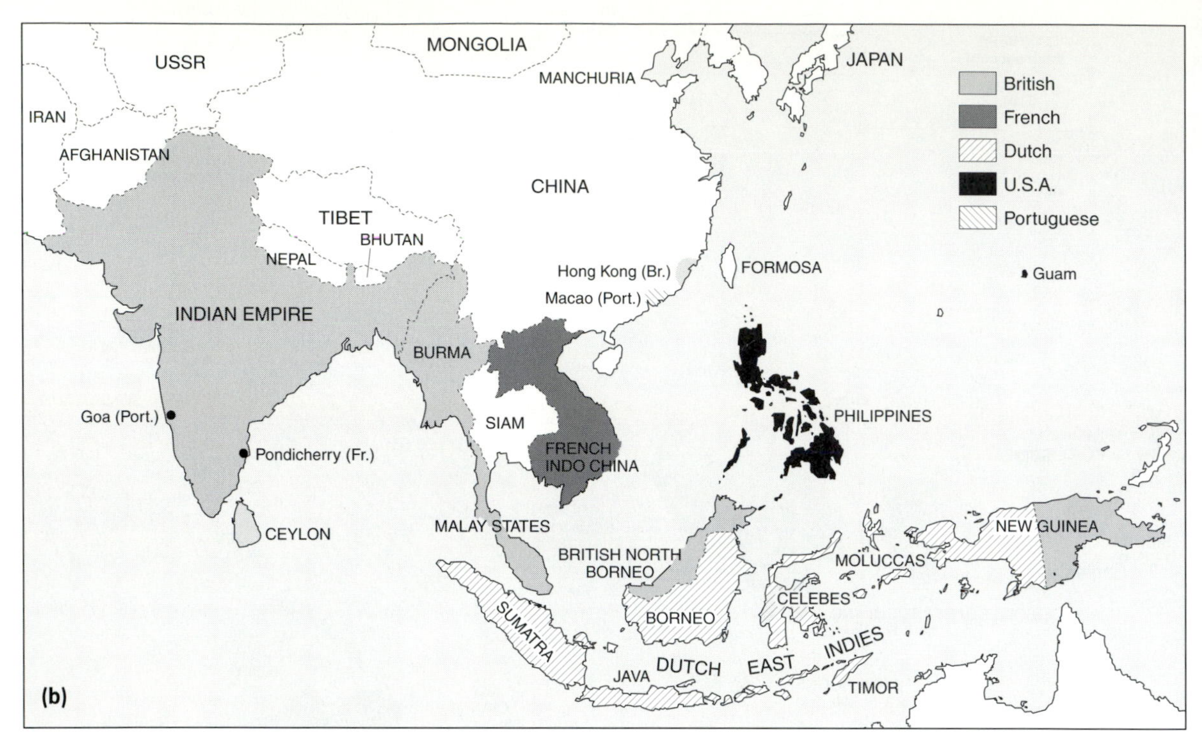

total political independence have in recent years been articulated among the Scots and Welsh in the United Kingdom, the Bretons and Basques in France, the Catalonians and Basques in Spain, the Flemish and French-speaking Walloons in Belgium, and in several other countries. In fact, some nationalist leaders have argued that the presence of the EU provides a rationale for independence. For example, the Scottish Nationalist Party's slogan is "Independence in Europe". Scottish Nationalist leaders argue that an independent Scotland, which nevertheless belongs to the European Union, would prosper more than a Scotland that remains part of the United Kingdom. Such arguments are rationalized on the grounds that several other members of the EU, for example, Denmark, Ireland and Finland, are smaller and weaker than Scotland itself.

Despite these difficulties, European unification continues to proceed at a pace undreamed of by world leaders during and after World War II. In 1963, political geographer Saul Cohen predicted Europe's geopolitical development with surprising accuracy:

> If Free Europe has a new destiny to fulfill, it will fulfill it through internal consolidation together with reinforced external associations. For European unity cannot be achieved behind economic and military frontiers. It can only be achieved along a new pathway, which will combine historic benefits of regional specialization stemming from global associations and the modern advantages of regional consolidations emerging from the breakdown of national differences (Cohen 1963).

Clearly, European history since 1963 has borne out the accuracy of Cohen's predictions. To an extent, Europe's position during the Cold War can be regarded as a vindication of French internationalism in geopolitics before World War II. Indeed, Europe's position in the global arena today is reminiscent of France's position between maritime Britain or the Rimland and continental Germany or the Heartland. Europe's perception of itself as a buffer zone between the United States and the Soviet Union is of fundamental importance in understanding its contemporary geopolitical position, both during and after the Cold War. Yet the end of the Cold War has generated a new set of geopolitical concerns for the traditional center of the global economy, as we shall see in greater detail in subsequent chapters.

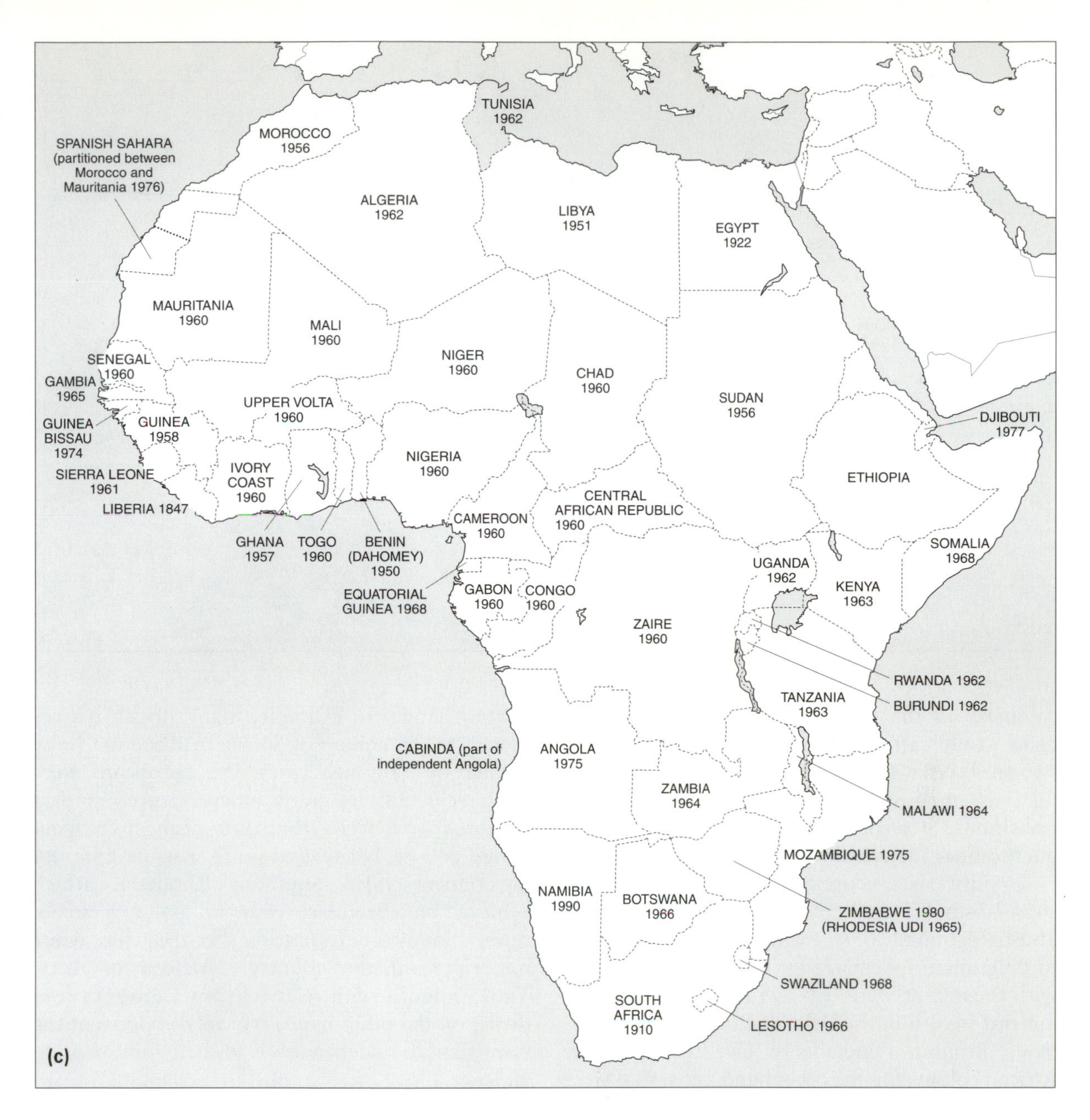

3.8 The Third World in the Cold War

The Cold War had profound effects on the world outside Europe and North America. In particular, the Cold War strongly influenced the history and development of the less developed former colonies of Africa, Asia and Latin America – the so-called "Third World".

At the beginning of the Cold War, a large majority of the world's population lived under European colonial rule. Following the end of World War II, independence sentiments became increasingly strident in many colonies. Within thirty years, nearly all achieved political independence (Figure 3.1a–d pp. 29–32).

In some of the former European colonies, the transition from colonial status to political independence was peaceful and non-violent. Other

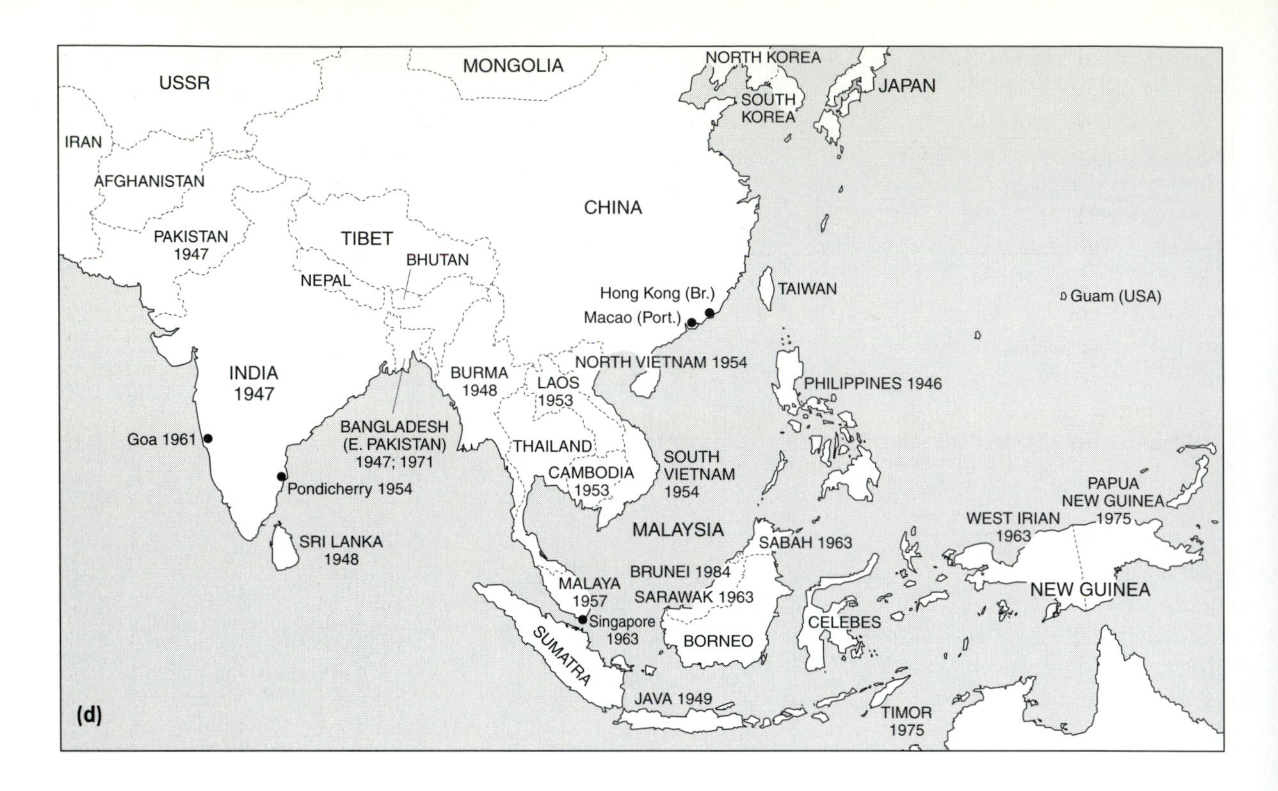

colonies, on the other hand, achieved independence only after long, violent revolutionary struggles. What factors are associated with peaceful, as opposed to violent, independence transitions? Recent scholarship addressing this question has identified several relevant factors.

In most cases, economic motivations underlay colonialism. Thus colonies characterized by unusual economic or strategic value were less likely to be granted independence quickly as compared with those that were less lucrative. For example, mineral wealth in the Belgian Congo (now Zaïre) made Belgium reluctant to give up its major African colony. On the other hand, those that were lacking in valuable resources tended to achieve independence without a struggle. Likewise, isolated or poorly accessible colonies were more likely to gain independence painlessly than those that were easily accessible or located strategically.

Ease of transition to independence was also related to the size of the colony's European population. Some colonies in the Third World attracted large numbers of European settlers. Typically, European settlers controlled a disproportionate share of the colony's resources, wealth and agricultural land. For example, many British natives moved to the colony of Southern Rhodesia (now Zimbabwe). In such cases, the Europeans were often reluctant to grant independence, in part because of concern that European investment would be lost. For several years, a white-minority government ruled Southern Rhodesia, which achieved independence only following a sometimes bloody confrontation between the white minority and the country's African majority. Those colonies with relatively few European residents, on the other hand, typically underwent the transition to independence with a minimum of violence.

The extent to which colonial residents were permitted or encouraged to participate actively in colonial government also influenced the violence of the independence struggle. Those colonies which contained a relatively large number of well-educated local residents were less likely to have to struggle violently for independence. To a considerable extent, this distinction is related to the fact that philosophies of colonialism and colonial administration differed considerably among the European colonial powers. Some tended to promote internal self-government. Education of the local population was encouraged, and talented local youths were encouraged to obtain European educations before obtaining positions of responsibility in the colony's army and civil service. In many cases, these Western-trained individuals succeeded to political leadership once their countries achieved independence. Colonies of countries which had encouraged self-government and education within their colonies, in particular Britain, tended to achieve independence peacefully. On the other hand, those colonies in which local participation and education were discouraged were more likely to experience violent independence struggles.

Once independence was granted, the newly independent countries of the Third World faced a variety of problems, many of which were the result of the process of colonialism itself. As we have seen, colonialism was the vehicle by which the less developed countries of the world were incorporated into the global world economy centered in Europe. Raw materials and labor from Europe's colonies were used to fuel European industries, and profits flowed steadily from the Third World to Europe. Colonialism rendered many former colonies extremely dependent on their former colonial masters in particular, and on the developed countries in general. By the time independence was granted, most former colonies lagged far behind Europe and North America in per capita income, education, employment and most other conventional measures of economic development.

In order to redress these imbalances, many of the newly independent former colonies sought assistance from the more developed countries of the world. Frequently, however, such assistance was tied to the Cold War world order. Both sides in the Cold War attempted to line up allies throughout the world, and various types of financial assistance, both military and non-military, were used in order to develop and maintain alliances between the less developed countries and the United States or the Soviet Union.

The superpowers, meanwhile, used various forms of foreign aid in order to promote their own foreign policy objectives during the Cold War era. Less developed countries which were located strategically or whose support was considered critical to American or Soviet military or political objectives tended to receive large quantities of assistance. For example, one recent study has examined the distribution of American food aid to less developed countries during the Cold War period. Few would doubt that the need for food aid was greatest in famine-stricken sub-Saharan African countries such as Sudan, Ethiopia and Somalia. Yet the four countries which received the most aid during this period were Poland, Yugoslavia, Israel and Egypt – relatively less developed countries to be sure, but hardly characterized by the levels of famine and starvation that we associate with more impoverished areas. Clearly, the magnitude of food aid provided by the United States was commensurate with American foreign policy objectives more than with immediate needs (Kodras 1992).

Many less developed countries, and especially those located in strategically important areas, found themselves courted by both superpowers during the Cold War. Soon their leaders realized that it was seldom in their interests to become too closely identified with either the Americans or the Soviets. In this way, they could obtain needed economic and military assistance from both sides. Thus many Third World leaders managed to increase their overall volume of foreign assistance by playing off the superpowers against each other. Over time, they would tend to "side" with the United States for a few years, with the USSR for a few years, then back to the United States for a few more years and so on.

3.9 Shatterbelts

As we have seen, many Europeans expressed concern that Europe would be the major battleground of the Cold War. In fact, such predictions proved false. Rather, the less developed countries of the world proved to be the battleground of the Cold War. During this period, millions of Third World residents were killed, injured or lost their homes and possessions in the numerous arms conflicts that took place in less developed countries during the Cold War period. Millions more became refugees, driven out of their native countries as a

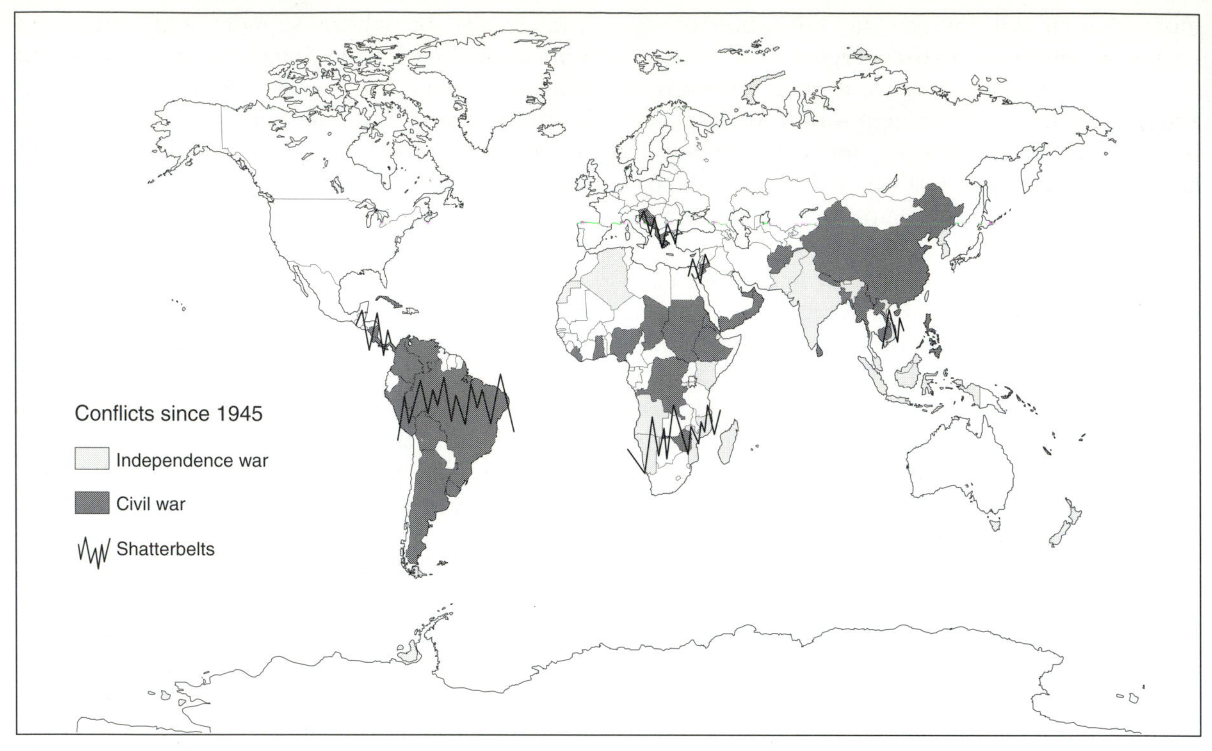

Figure 3.2 Distribution of shatterbelts

result of political or religious persecution.

During the Cold War, hundreds of wars were fought throughout the less developed countries of the world (Figure 3.2). Yet hostilities during the Cold War were concentrated especially in a few critical regions known as shatterbelts (see also Chapter 5). Four major shatterbelts of the Cold War have been identified: the Middle East and southwest Asia, Southeast Asia, southern Africa and Southeastern Europe (Cohen 1973). These areas have been identified as shatterbelts for two reasons. First, each area was of major strategic importance to the two superpowers. In addition, each was characterized by long-standing rivalry between indigenous ethnic groups with distinctive and often incompatible goals. For this reason, borders between states within shatterbelts have frequently been subject to instability and change.

The Middle East, along with neighboring North Africa and Southwest Asia, has been a source of substantial international conflict throughout the Cold War. The Middle East stands as a major crossroads between the three major continents of the Old World. Moreover, its vast petroleum deposits have proven critical to the economies of industrialized nations throughout the world. Long-standing ethnic tension within the area was exacerbated by the movement of European Jews to Palestine and the subsequent establishment of the state of Israel in 1948. Several major wars between Israel and hostile Arab neighbors have broken out over the years. Indeed, the last Cold War conflict – or perhaps the first post-Cold War engagement – was the war between the United Nations and Iraq over Kuwait in 1991.

Southeast Asia, like the Middle East, contains a large number of ethnic groups with different religious views and incompatible national objectives. The area is located along the major trade route between East Asia and South Asia, the Middle East and Europe, thus guaranteeing the region substantial strategic importance to both sides in the Cold War. As in the case of the Middle East, the region has been affected by colonialism and substantial migration from outside the region in recent centuries. Immigrants from China and India play important roles in the business and professional communities and the political lives of several Southeast Asian countries, exacerbating tensions between local ethnic groups.

South Africa's strategic importance stems from its vast deposits of minerals, including some found nowhere else in the world, along with its position along the main oceanic trade route between Europe and Asia. In addition, conflict in South Africa has stemmed from tensions between the descendants of European settlers who have moved to the area within the last three centuries. South Africa, which imposed a rigid system of discrimination against non-Europeans known as apartheid, was particularly prone to conflict during the Cold War. It is to be hoped that the elimination of apartheid and the subsequent election of an African-majority government in 1994 will ease tensions in this troubled region of the world.

Ethnic tensions in the fourth major shatterbelt, southeastern Europe, have been brought to the world's attention since the end of the Cold War with the bloodshed in the former Yugoslavia. Yet this region was an important battleground of the Cold War. It was in Greece and Turkey that the Truman Doctrine was first formulated; while, as we have seen, the United States donated substantial amounts of aid to ethnically divided Yugoslavia during the Cold War in order to encourage Yugoslavia's independent, anti-Soviet foreign policies. Not only was southeast Europe an important shatterbelt during the beginning of the Cold War, but it appears that this region, like the Middle East, is likely to remain a venue of bloodshed during the post-Cold War era.

3.10 Non-alignment

While some leaders of the less developed countries were quick to play the Soviets and the Americans against each other during the Cold War, others found the Cold War incompatible with their basic objectives of promoting development. Some Third World leaders explicitly rejected the Cold War and emphatically refused to take sides with either East or West. These leaders developed an alternative approach known as non-alignment.

Those adhering to the principle of non-alignment rejected explicitly the idea that the world should be divided into two hostile camps. Not only did the philosophy of non-alignment reject both sides in the Cold War, but its leaders pointed to the concentration of poverty and warfare in the less developed countries during the Cold War era as evidence that the Cold War had reinforced the long-standing gaps in standards of living between the developed and the less developed countries. Thus leaders of the non-alignment movement argued that less developed countries, many of which were newly independent former European colonies, should work together to promote the economic and political interests of the Third World.

The philosophy of non-alignment soon became linked with the New International Economic Order, discussed in greater detail in subsequent chapters. Hence non-alignment came to be associated with efforts to reduce conflict between East and West, replacing this with a worldview emphasizing the structural economic differences between North and South. The North–South perspective on geopolitics has become increasingly important in light of the end of the Cold War, as we will see in Chapter 4.

3.11 The end of the Cold War

After more than 40 years, the Cold War came to a sudden and surprising end in the late 1980s and early 1990s. As we have seen, Mikhail Gorbachev's Soviet government promoted more liberal and less dogmatic domestic and foreign policies than its predecessors. The Gorbachev government lifted restrictions on trade and emigration, promoted cultural exchanges and permitted increased interaction between the Soviet satellites of Eastern Europe and the West.

In 1989, however, communist domination of Eastern Europe came to an abrupt end. Within less than a year, communist regimes in Poland, Czechoslovakia, Hungary and East Germany were replaced by elected democratic governments. In Poland, for example, Lech Walesa, an electrician and the leader of the Solidarity movement, became the first freely elected president of that country since World War II. Once the Communist government of East Germany collapsed, leaders of both

Plate 3.4 Voting in the city of Yaroslavl, Russia, during the referendum on March 17, 1991 about keeping the USSR together as a federation (photo: Kathleen Braden)

East and West Germany moved swiftly to promote reunification. By November 1990, German reunification had been completed.

The collapse of communism soon spread to the Soviet Union itself. As we saw in the previous chapter, the Soviet Union included Russia plus fourteen Soviet republics, which formed a collective buffer zone around the Russian core to the west and south. Once the communist governments of Eastern Europe had collapsed, efforts to promote increased liberalization arose within the Soviet Union itself. Anti-Soviet sentiment was especially strong in the Baltic republics of Lithuania, Estonia and Latvia. Lithuania in fact declared its independence from the Soviet Union in late 1989, although its independence was not recognized by Gorbachev's government.

In August 1991, hard-line opponents of Gorbachev's regime attempted a *coup d'état* against the government. The coup effort failed, but Gorbachev stepped down. By the end of the year, the Soviet Communist Party had voted itself out of existence. Russia, which includes half of the Soviet population and three-quarters of its land area, established a democratic government with Boris Yeltsin, who had played a key role in preventing the hard-line coup attempt, elected to the Presidency. Similar efforts at democracy were attempted in the former Soviet republics. In early 1992, eleven of them (all but the Baltic republics and Georgia) formed an economic union known as the Commonwealth of Independent States. Soviet communism had faded into the pages of history, and the Cold War was now over.

4

Geopolitics And The Less Developed Countries

KEYWORDS

development,
demographic transition,
economic sectors,
colonialism,
dependency,
non-alignment,
New International Economic Order

KEY PROPOSITIONS

- Have less developed states been major actors in geopolitics?
- How has colonialism affected development?
- How are economic structure, development, and population change interrelated?
- How may a "New International Economic Order" influence less developed countries?

4.1 Introduction

Formal geopolitical theories have been articulated throughout the twentieth century. For the most part, as we have seen, these theories have focused on the developed countries of the world. Despite the fact that a large majority of the world's people live outside the developed countries, the less developed countries of the world have received relatively little attention from geopolitical thinkers.

Not only have formal geopolitics tended to ignore the less developed countries, but even ordinary Americans tend to ignore them. For example, college students at three American universities were recently asked to rank the 55 countries of the world that have populations of more than 15 million on the basis of "geopolitical importance" (Archer *et al.* 1997). The students ranked the United States, the large countries of Western Europe, Russia, China and Japan as very important. Most of the less developed countries, by contrast, were regarded as unimportant. Ranked near the bottom in order of importance were countries such as Mozambique, Sri Lanka, Uganda and Burma.

The purpose of this chapter is to consider relationships between geopolitics and the less developed countries of the world. Why have geopolitical theorists paid so little attention to these countries and their populations? How do residents of the less developed countries themselves view geopolitics? Is the geopolitical position of the less developed countries likely to change now that the Cold War is over? In this chapter, we first consider the geopolitical position of the so-called "Third World". We then discuss how geopolitical theory has treated the less developed countries. In doing so, we examine the issue of geopolitics and the less developed countries from the point of view of traditional Western geopolitics, and then from the point of view of the less developed countries themselves. Finally, we address how the end of the Cold War may be affecting the geopolitical position of less developed countries.

4.2 The less developed countries today

A large majority of the world's people live outside Europe and North America. For the most part, however, geopolitics has paid little attention to places outside the core of the world economy. In the past, of course, few of the less developed countries were politically independent. Most now-independent "Third World" countries had at one time been under the formal political control of one or more of the Western powers. As recently as 1950, only four countries on the entire continent of Africa (Egypt, Ethiopia, Liberia and South Africa) were politically independent. The others, accounting for more than 80 per cent of Africa's population, were European colonies.

Today, nearly all of Europe's former colonies are politically independent. An Africa that contained only four independent countries half a century ago, today includes over 40 independent, sovereign states. Indeed, all of Africa's people now reside in independent countries. Yet political independence has not always translated to economic development. Over the past several decades, the gap in standards of living between the less developed countries and the advanced industrial economies of Europe, North America and Japan has increased. Despite decades of efforts to promote economic development in Asia, Africa and Latin America, these regions lag far behind the core countries in almost any measure of development.

Maps comparing basic social indicators illustrate the dramatic contrasts in the overall quality of life between the less developed and developed countries. Per capita annual income, for example, ranges from less than $200 in some of the poorer nations of Africa to over $10,000 in the United States and several European countries (Figure 4.1). Access to education, taken for granted in the West, is restricted in many less developed countries where millions of children lack the opportunity for even basic schooling (Figure 4.2). Illiteracy remains a major social problem in less developed countries throughout the world.

Health standards also differ substantially. Although the gap between the developed and the underdeveloped countries has narrowed in recent years, life expectancy in the less developed countries is 20 or more years less than is the case in the United States (Figure 4.3). To a considerable

Figure 4.1 World gross national product. Source: 1997 World Population Data Sheet, Population Reference Bureau

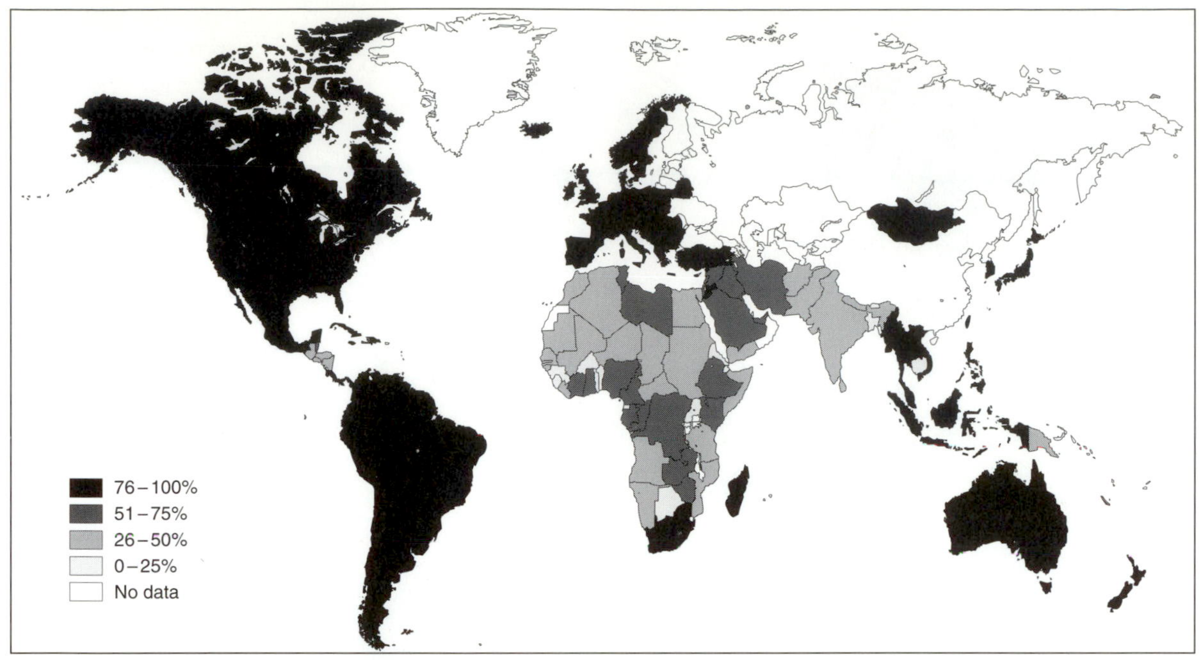

Figure 4.2 World literacy. Source: 1997 World Population Data Sheet, Population Reference Bureau

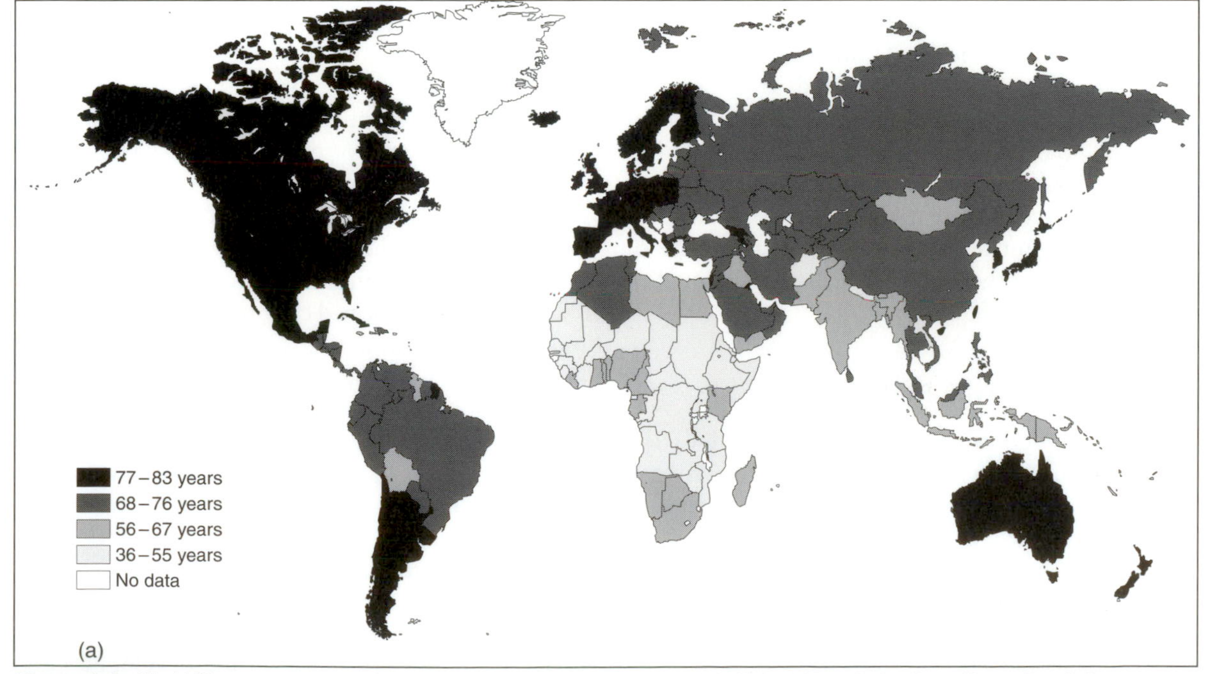

Figure 4.3 World life expectancy: (a) female; (b) male (on p. 40). Source: 1997 World Population Data Sheet, Population Reference Bureau

Figure 4.3 Continued

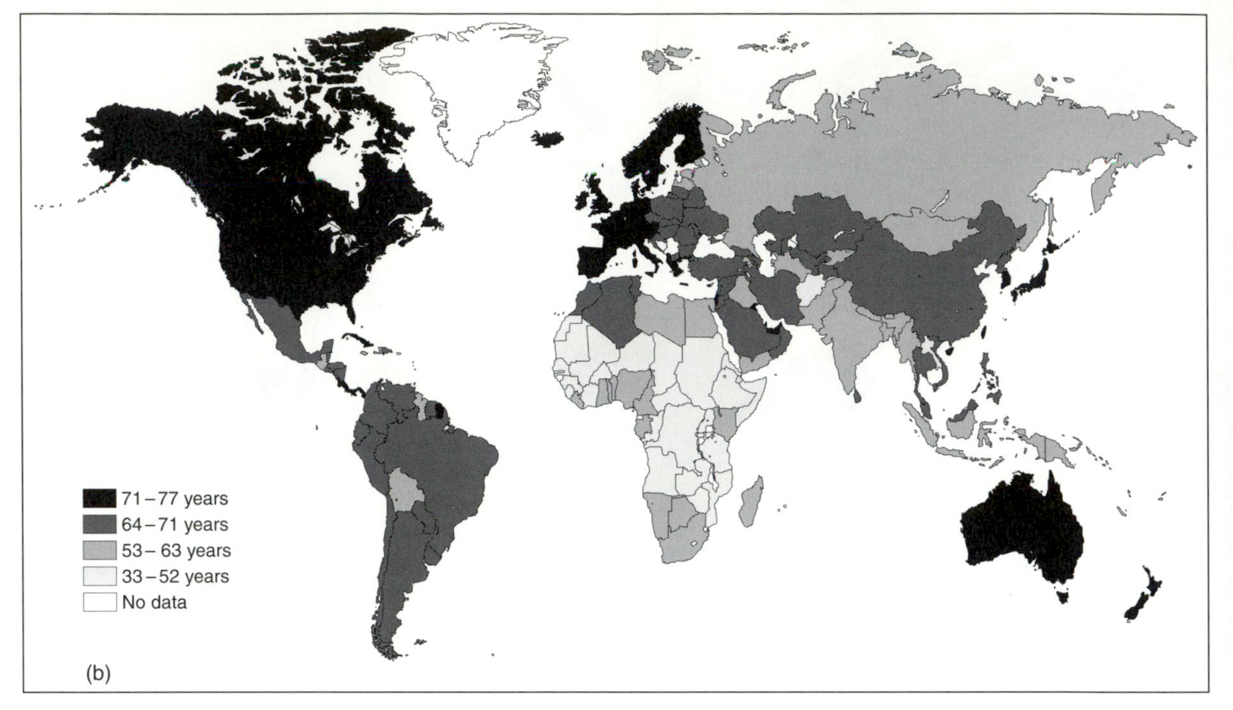

extent, the large differences between developed and underdeveloped countries in health care quality result from a lack of effective medical care in the less developed countries. Access to even basic medical services, such as immunization against infectious diseases, first aid and treatment of common childhood diseases, is absent or restricted in many parts of the less developed world. Governments in less developed countries lack sufficient resources to treat disease, and many less developed countries lack sufficient numbers of physicians and other trained medical personnel to combat disease adequately.

4.2.1 DEMOGRAPHIC TRANSITION

How can the contrasts in health care, education and other measures of the quality of life between developed and less developed countries be explained? Three basic contrasts can be articulated. These include contrasts in demography and economic structure.

One of the most fundamental distinctions between the developed and the less developed countries involves rates of population growth. Although population growth rates in some less developed countries have declined substantially in recent years, population growth rates are considerably higher than they are in the developed countries of the world. Today, in fact, over 90 per cent of babies are born in less developed countries. Large numbers of newborn babies are adding to already dense populations in some countries. Officials in densely populated less developed countries such as Bangladesh express considerable concern that the rapid rate of population growth will result in famine, shortages of basic resources and a general decline in the standard of living.

Many less developed countries are characterized by young populations with high birth rates. Lowered life expectancy and higher birth rates result in much younger populations in less developed countries. In the United States, the median age of the population is 31 years, with elderly persons over 65 outnumbering teenagers between 13 and 19. Mexico, on the other hand, has far more teenagers than elderly persons, and its median age is barely 18.

The demographic transition model accounts for some of the striking differences in demography between the developed and underdeveloped parts of the world (Figure 4.4). Based on empirical evidence from the advanced industrial societies, the

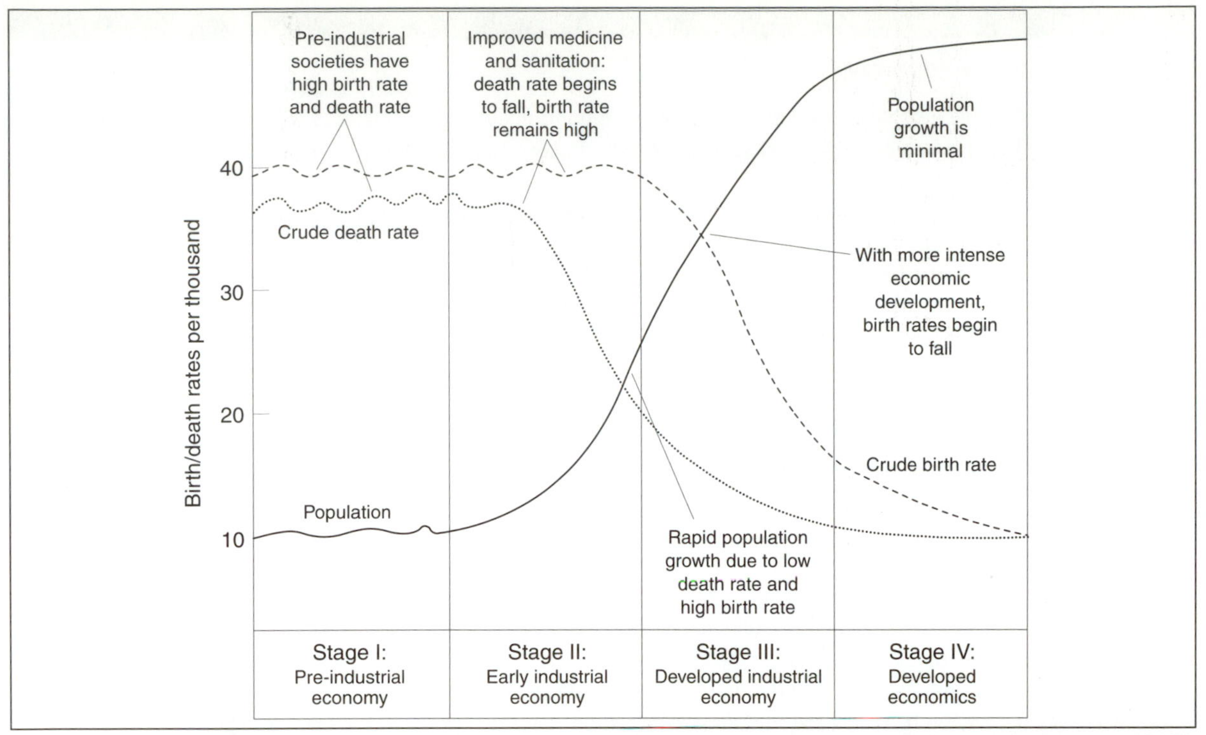

Figure 4.4 The demographic transition model (from R Potter et al, *Geographies of Development*, (Addison Wesley Longman,1999))

demographic transition model posits that a society's death and birth rates change over time in a predictable fashion. In pre-industrial societies, birth and death rates were both high – much higher than is characteristic of the world today. Death rates fluctuated around high means, and famines and epidemics, which killed large numbers of people, caused dramatic short-run increases in death rates. This first stage of the demographic transition was characteristic of medieval Europe and of the less developed countries of the world prior to European contact.

Industrialization and economic development resulted in the widespread diffusion of more advanced medical technology. This in turn led to a rapid decline in the death rate. Modern medical technology reduced infant mortality, eliminated most major epidemics and resulted in the control of infectious diseases such as diphtheria, smallpox and cholera. Public health measures were enacted, promoting improved sanitation and reducing the spread of disease. These declines in the death rate, which characterized the second stage of the demographic transition, occurred during the eighteenth and nineteenth centuries in the developed countries.

While death rates decline dramatically during the second stage of the demographic transition, birth rates remain high. There are several reasons why birth rates in the less developed countries remain substantially higher than those in Europe and North America. For residents of many less developed countries, children are seen as a form of social security. A couple raising children expect that their grown children will support them in their old age. In societies lacking government social insurance programs typical of the West, support from children is often the only insurance against poverty and starvation in old age. Moreover, high rates of infant mortality characterized pre-industrial society. Although infant mortality is much lower in the less developed countries today than was the case several decades ago, rates of infant mortality remain much higher than they are in Europe and North America. With high rates of infant mortality, raising large families helps to ensure that several children will survive to adulthood.

BOX 4.1

Tourism and development

In recent years, many less developed countries have begun to encourage tourism as a means of promoting development. Tourism has brought needed revenues into many countries, but has been a cause of significant problems as well.

Tourism is travel to places outside everyday experience undertaken for recreational purposes. During the twentieth century, tourism has increased dramatically throughout the world. Increased leisure time, increased disposable income and improved transportation encouraged people to undertake tourist travel. By the 1960s, international tourism had become increasingly popular. At first, most international tourists traveled to Europe. Today, international tourism is globalized.

Steady growth in the tourist industry has induced many countries to encourage the development of tourism as an economic base. Tourism can expand a country's revenue base; it can also create cultural conflict, undermine the local economy and cause large-scale environmental degradation. The economic benefits of tourism are dampened by the fact that many hotels, restaurants, and other facilities are owned by multinational corporations or by individual Europeans and Americans. Many local people hold low-paying, menial jobs.

Nature and heritage tourism is becoming more and more popular. American, European, and other tourists pay increasingly large sums of money to view exotic wildlife and important cultural landscapes and resources. The maintenance of habitat for large endangered animals has caused controversy in some host countries. Small farmers living nearby have objected not only to the increased land values associated with taking potentially arable land out of production, but also to the presence of large numbers of dangerous animals. For example, Nepalese farmers living near the boundaries of Nepal's large reserve for endangered wildlife have pointed out that the tiger population has increased within and outside of the reserve and have blamed tigers for killing their livestock as well as attacking human beings.

Respect for local cultural norms is another issue associated with the international tourist industry. For example, residents of Muslim countries object to the presence of Western tourists who wear skimpy bathing suits and drink alcoholic beverages – practices that many Muslims find offensive. Tourism can also cause ecological damage. In several Caribbean islands, for example, tourism has resulted in severe ecological disruption because so many tourists have cut and removed coral from coral reefs, overfished, caused air and water pollution and otherwise damaged the fragile island ecosystems. If damage becomes too serious, tourists eventually refuse to visit. In some places, tourism has been all but abandoned, but the presence of tourism has placed permanent strains on local cultures, resources, ecosystems and economies. What was originally intended as a boost to development has ended up as an ecological and economic disaster.

Another important consideration underlying high birth rates in the less developed countries is the narrow range of opportunities available to women. In many less developed societies, early marriage and childbearing remains a social norm. Considerable social pressure is placed on women to marry and bear children at an early age. The absence of education and employment opportunities reduces the set of alternatives available to women in the less developed countries. In some countries access to birth control and family planning advice is lacking. Moreover, religious values encourage high birth rates. Some religious traditions prohibit contraception, and some encourage women to raise large families.

At the third stage of demographic transition, birth rates begin to fall, eventually reaching levels commensurate with the now low death rates. This third stage of demographic transition typifies the advanced industrial societies of the world today. In developed countries, families are often small and early marriage and childbearing are less usual. Careers and higher education provide women with alternatives to the raising of large families. An increasing percentage of women delay or forego childbearing, thus further retarding rates of population growth.

The demographic transition model posits that rates of population growth in the less developed countries will eventually decrease to rates comparable to those of developed countries today. Whether this will actually take place, of course, remains to be seen. Much depends on the extent to which rapid population growth is viewed as detrimental to society – a question of considerable controversy. Many experts argue that rapid rates of population growth are resulting in significant resource depletion. This view is consistent with a theory of the relationship between population and resources set forth two centuries ago by the British philosopher and economist Thomas Malthus, who regarded the natural resource base as fixed, and consequently argued that population increases would result in resource depletion. The result of resource depletion would be famine, war and disease.

In contrast, other observers tend towards a more optimistic view of the relationship between population and resources. This so-called "cornucopian" view reflects the belief that increasing populations cause a society's resource base to expand, therefore creating an increase in resources per capita. This expansion results from the possibility that larger populations permit increased specialization, an improved and more efficient division of labor, and the opportunity to expand economic surplus.

These contrasting views underlie various policies intended to address the issue of population growth in different countries. Because the decision to bear a child is a very personal and intimate one, many such policies have proven quite controversial. In some countries, government has taken an active role in lowering birth rates. China's controversial population policy, which encourages families to have only one child, typifies a Malthusian approach to the relationship between population and resources. The Chinese one-child policy proved so controversial and unpopular, however, that in recent years the Chinese government has relaxed it.

Prior to the end of the Cold War, on the other hand, the government of the Soviet Union and some of its Eastern European satellites adopted a cornucopian view and encouraged high rates of fertility. In order to encourage higher birth rates, the Soviet government rewarded mothers who bore large numbers of children. Once the Soviet government collapsed, these incentives were eliminated by the new democratic governments in Russia and the former Soviet republics. Today, birth rates in Russia and Eastern Europe are among the lowest in the world.

4.2.2 ECONOMIC STRUCTURE IN DEVELOPED AND LESS DEVELOPED COUNTRIES

The developed and underdeveloped countries of the world also differ significantly in economic structure. Economists classify stages of an economic production process into the primary, secondary, tertiary and quaternary stages of production. The primary sector involves the extraction of raw materials and includes agriculture, mining, fishing, forestry and similar occupations. The secondary sector involves the manufacturing of finished products. The tertiary sector, sometimes called the service sector, includes activities related to the distribution and servicing of the finished products. The quaternary sector includes government, education, research and development and information-related activities.

Relative to the developed countries, less developed countries are heavily dependent on the primary and secondary sectors of the economy. In the United States, for example, only slightly more than 1 per cent of the population work as farmers. Nevertheless, American agriculture is so efficient that American farmers provide enough food for the entire nation and a considerable surplus for export abroad. Thus the average American farmer produces enough food annually to feed over 100 people. In contrast, agriculture remains the most common occupation among residents of many underdeveloped countries. Although the percentage of persons involved in other occupations has been increasing, in some less developed countries over half the population remains in agriculture.

The percentage of workers employed in the secondary sector is also declining in the United States and Western Europe, while it is increasing in the less developed countries. Today about 18 per cent of Americans work in the manufacturing sector. Thus about 80 per cent of all jobs in the United States are in the tertiary and quaternary sectors of the economy. In contrast, the tertiary and quaternary sectors of the economy of most less developed countries are poorly developed.

This disparity is exacerbated by the fact that in many countries tertiary and quaternary jobs are dominated by Western-controlled multinational corporations. Many multinational corporations divide their production activities between developed and underdeveloped countries. Tertiary and quaternary activities along with high-wage, high-skill secondary production are concentrated in the developed countries; while low-wage jobs requiring a minimum of skill are found in less developed countries, where labor is cheaper and wages are lower. The result of this new international division of labor is that the economies of the underdeveloped countries become increasingly dependent on the activities of multinational corporations, most of which are headquartered in the developed countries.

The disparities in income, wealth and employment opportunities between the developed and underdeveloped countries are reflected also in the considerable volume of temporary and permanent migration between underdeveloped origins and industrialized destinations. Throughout northern and western Europe, for example, a large proportion of menial labor is undertaken by workers of foreign origin. Turks, Greeks, Arabs and natives of the former Yugoslavia have moved to Germany and Switzerland; while the United Kingdom is home to an increasing number of natives of former British colonies including India, Pakistan and the Caribbean islands. A similar role is played by many Mexican nationals living and working in the United States. Although conditions for these immigrant "guest workers" are often substandard, many nevertheless find menial employment abroad preferable to marginal economic conditions at home. Although many migrants send much of their earnings home, the incentive to migrate to developed countries robs underdeveloped nations of considerable labor resources.

By no means are all migrants from the less developed to the developed countries unskilled or uneducated. Many are refugees; a smaller but influential number of migrants from less developed countries to the United States and Western Europe are highly educated physicians, engineers, scientists and business executives. Western-educated natives of less developed countries often choose to practice their professions in developed countries, where they can earn much higher salaries and in many cases avoid the possibility of persecution from repressive governments. At a time when many less developed countries are suffering from a substantial shortage of trained and qualified doctors and other professionals, this "brain drain" has had the effect of retarding development in many parts of the world.

4.3 Colonialism and the less developed countries

The previous discussion illustrates the extent to which the less developed countries remain politically and economically dependent on the advanced industrial societies, and as a result remain relatively poor and underdeveloped. Poverty, underdevelopment, overpopulation, environmental degradation and resource depletion characterize many less developed countries today. The relationship between the developed and developing worlds has evolved through a complex series of historical forces over the past 500 years. To a large extent, the history of the Third World is one of the effects of the expansion of the European-centered global economy to encompass the entire planet. This history involves the systematic incorporation of areas outside Europe into the world economy – an incorporation process often associated with changes in political control and accompanied by violent conflict.

The process of systematic incorporation began in the late Middle Ages, when the capitalist world economy developed in Western Europe. During the Age of Exploration, European explorers discovered and mapped the previously unknown areas of the world. Exploration was soon followed by colonization. Early efforts at colonialism generally involved attempts to exploit specific resources in the newly discovered lands. Areas that contained precious metals and spices, or that were believed to contain them, were plundered, with huge surpluses transferred to Europe. Initially, South America was a prime target for colonization on account of its vast supplies of gold and silver. Precious metals valued at millions of dollars were mined and transported to fuel the

European economy. It has been estimated that by 1550 – barely half a century after the first voyage of Columbus – as much as one-seventh of the adult male population of Peru worked in the silver mines at Potosi.

4.3.1 COLONIALISM AND AGRICULTURE

The early years of colonialism also revolutionized global agriculture. Prior to the colonial era, most areas throughout the world were self-sufficient in agricultural production, and most crops were consumed locally. The age of colonialism resulted in the elimination of self-sufficiency in most areas of the world. Instead, the European world economy assumed control of most aspects of the world's agriculture, with especially significant effects in the less developed countries.

In the less developed countries, a considerable proportion of agricultural production became concentrated onto plantations, which in many areas replaced small subsistence farms. Plantation agriculture affected the relationship between a society and its land resource base. No longer were farmers producing crops for local consumption. Instead, plantations were worked by slave or wage laborers, whose produce was exported. Only a small minority of the population owned or controlled most or all of the land, and plantation owners were often absentee landlords whose primary concerns were European rather than indigenous.

The alienation of agricultural workers from the land was exacerbated by large-scale labor transfers, such as those associated with the slave trade and by the transfer of crops to different areas of the world. Demands for the exploitation of precious metals and other resources resulted in great demand for labor. From the fifteenth to the early nineteenth centuries, millions of Africans were captured and transported to the Americas as slaves. African and Indian slaves worked on plantations and in mines throughout the Americas, producing huge surpluses for European consumption. Today, the European colonial powers' establishment of large plantations continues to affect agriculture in the Third World. Entire plantations and indeed entire regions are devoted to the cultivation of a single crop – a practice known as monoculture. Because of monoculture, the economy of an entire country can depend on the performance of only one or two agricultural products. In such cases, the country's economic performance can hinge on world market prices for commodities far beyond its control.

Cuba provides an interesting case in point. The Cuban economy has long been dependent on the production of sugar cane for export markets. In recent years, the world market price of sugar and sugar products has tended to decline as a result of medical evidence that sugar consumption can be harmful to health. This evidence has also inspired the development and production of sugar substitutes, further weakening demand for the product. The decline in the world sugar market in conjunction with Cuba's dependence on this product has had disastrous consequences for the island's economy.

In addition to its economic consequences, monoculture renders a country dependent on favorable natural conditions. Floods, pests, storms and drought can wipe out crops in an entire region. When the region is largely dependent on a single crop, these natural disasters can ruin a regional economy. In contrast, the effects of natural disasters are mitigated when several crops are grown on a single farm or within a region. Even if one crop were to fail, others would grow successfully, assuring the farmer of at least a minimum profit.

4.3.2 COLONIALISM AND THE POLITICAL AND ECONOMIC ORGANIZATION OF COLONIES

The European powers organized their colonies in order to maximize European profits, with far-reaching consequences for local economies. As one of the primary purposes of colonialism was the establishment of trade linkages between the colony and the European mother country, the Europeans organized local transportation and urban development to suit their needs. The Europeans concentrated urban development in seaports, which served as conduits between the colony and the mother country. Many such seaports expanded to become the major cities in less developed countries today.

In locating these cities, European powers were often anxious to avoid conflict with powerful indigenous peoples. Hence they took care to establish their new cities at sites distant from existing power centers. Because they deliberately avoided locating cities near centers of indigenous power, many such cities were established on swampy, unproductive and unhealthy territory. Cities such as Lagos in Nigeria and Calcutta in India were located on what had been swampland.

In developing infrastructure, the European colonial powers concentrated on those developments that maximized economic returns to themselves. Thus internal transportation in European colonies tended to focus on the port cities. Railroads and roads were built to connect the interior with the port, but transportation between ports within the interior was often difficult and in many places remains so today. The relative neglect of internal transportation was a particularly significant problem in those areas of the Third World in which the colonial powers drew boundaries that split indigenous populations. The drawing of colonial boundaries without regard to local cultural considerations was especially typical of Africa, which was divided into colonies following arbitrary agreements at European conference tables. Most of Africa came under European colonial rule between 1880 and 1914. At the beginning of this period, only a few outposts, located primarily along the coast, were under direct European control. Thirty-five years later, nearly the entire continent was made up of European colonies.

The establishment of these colonies divided many tribes between colonies of two or more European nations, and individual colonies generally comprised two or more indigenous cultures, which often had histories of conflict. The mismatch between the locations of indigenous cultures and those of colonial boundaries persisted after the colonies were granted political independence following World War II. Today, national boundaries in the Third World frequently cross traditional tribal boundaries. As in contemporary Europe, nationalist sentiment within many Third World countries has become an important foreign policy issue.

In general, the process of colonialism sowed the seeds of Third World dependency on the developed countries of the world. Not only did much of the Third World become subject to direct European political control, but its economy developed in such a way as to create and perpetuate economic dependency. This dependency underlies contemporary geopolitical theories as applied to less developed countries.

4.3.3 COLONIALISM AND INDEPENDENCE

During the twentieth century, revolutionary and nationalist movements arose throughout Africa, Asia and other European colonies. These grew steadily in strength with the passage of time, and after World War II demands for political independence escalated rapidly. The British colonies of the Indian subcontinent were granted independence in 1947, but partitioned along religious lines into the separate countries of India and Pakistan. In the late 1950s and in the 1960s, most of Europe's African colonies were granted independence. By

Plate 4.1 Poster for Washington State potatoes sold as a delicacy in Hong Kong (photo: Kathleen Braden)

1980 few colonies remained in the world, although economic dependency was seldom lessened in conjunction with political independence.

In some countries, the transition from colonial rule to independence was achieved with a minimum of violent conflict, whereas in other cases independence was achieved only following armed struggle. In general, those colonies that achieved independence without bloodshed were those which had encouraged at least some local participation in government and had encouraged Western education for local residents. In addition, a lack of an armed struggle has been correlated with high proportions of local residents in the army and in governmental bureaucracies. Those colonies in which large numbers of local residents were encouraged to join the civil service or the armed services and go into teaching and other professions were more likely to achieve political independence peacefully. In contrast, countries ruled with an iron hand by the Europeans were more likely to suffer violence in their quest for independence.

4.3.4 INTERNATIONAL TRADE

Controversy over these principles has extended to other aspects of international management of global trade. Over the course of the twentieth century, the volume of world trade has increased dramatically. Increased volumes of international trade are the result of several factors. Improved transportation and communications technology has made it far easier and cheaper to move goods over long distances. Meanwhile, countries have become less inclined to restrict or limit trade with other states.

Throughout recent world history, countries have been faced with the choice of whether to encourage or discourage unrestricted international trade. Those choosing to restrict trade have often done so by enacting tariffs. Tariffs are taxes imposed upon the movement of goods and services across international boundaries. Although countries sometimes justify tariffs on the grounds that taxes on imported goods encourage consumers to purchase domestic products and therefore help domestic industries, countries often react to tariffs on the goods that they produce by enacting tariffs of their own. Such "trade wars" have the effect of increasing the price of goods to consumers throughout the world, decreasing consumption and therefore slowing the pace of the economy.

During the early twentieth century, many countries enacted high protective tariffs. The United States, for example, enacted the Smoot-Hawley Tariff in 1929. The Smoot-Hawley Tariff imposed a 40 per cent duty on the import of many manufactured goods. In response, several other countries imposed similarly high tariffs. In the long run, these high tariffs depressed consumption, reduced production and therefore were widely blamed for the worldwide economic slowdown that led to the Great Depression.

After World War II, members of the United Nations expressed concern that high tariffs in the future might lead to another worldwide

Plate 4.2 Israeli vegetables for sale at an outdoor market in Helsinki, Finland (photo: Kathleen Braden)

BOX 4.2

GATT and international trade policy

Because the world's economy is characterized by global economic interdependence and division of the earth's land surface into independent sovereign states, national officials have debated the relative merits of encouraging or discouraging trade across international boundaries for centuries. Protectionist policies prohibit international trade or impose tariffs, or taxes, on the import or export of goods, in order to discourage cross-border trade. Free-trade policies, on the other hand, encourage trade across international boundaries.

During the 1920s and 1930s, many of the leading industrialized countries of the world imposed substantial protective tariffs. These high tariffs had the effect of reducing international trade, and many economists blamed the effects of the Great Depression of the 1930s on high tariffs in the United States and other countries. After World War II, many Allied leaders felt that a return to prosperity depended on the promotion of free trade between countries. In 1947, 23 countries signed the General Agreement on Tariffs and Trade (GATT). The purpose of GATT was to encourage and direct the expansion of international trade. Since its inception, more than 100 countries have signed the GATT agreement.

Strictly speaking, GATT is not a single treaty. Rather, GATT is a collection of hundreds of individual bilateral trade treaties among its members. The basic principle underlying these GATT treaties is the most favored national principle, which requires that each country must grant any more-favorable terms of trade granted to a third party to each other also. For example, if Brazil and France negotiate particular terms of trade, Brazil must grant France more favorable terms if it also grants these terms to another country. GATT maintains offices in Geneva, Switzerland, and is formally linked to the United Nations. Today over 85 per cent of international trade is conducted under the auspices of GATT.

GATT's members meet regularly to consider revising and updating GATT procedures. In 1986, the "Uruguay Round" of conferences began with a meeting in Punta del Este, Uruguay. The Uruguay Round addressed several deficiencies in the earlier GATT agreements. For example, earlier GATT agreements covered most manufactured goods, but did not cover agricultural products, services and intellectual property such as patents and copyrights. The Uruguay Round addressed these types of trade, and dealt with the increasingly important role of transnational corporations in international trade.

Do GATT agreements benefit less developed countries, or do they hinder development efforts? Experts in both the developed and less developed countries disagree sharply on this question. Proponents of GATT argue that the elimination of tariffs increases the overall volume of international trade, allowing larger quantities of less expensive foreign goods into less developed countries. On the other hand, critics have regarded GATT as a "rich man's club", arguing that GATT agreements reinforce the control of the developed countries over international trade and thereby preventing less developed countries from reducing their dependence on multinational corporations and on developed countries.

depression. Consequently, the United Nations established the General Agreement on Trade and Tariffs (GATT). GATT agreements are bilateral agreements between countries which discourage the imposition of trade barriers through the concept of "most favored nation" status. Any country that is granted most favored nation status must be granted terms of trade which are at least as favorable as those given others. For example, if the United States gives most favored nation status to Canada, it cannot offer a third country more favorable terms of trade than it offers to Canada.

The impact of this principle has been to reduce trade barriers worldwide, with a resulting explosion in the volume of international trade. Today, GATT agreements cover nearly 90 per cent of all international trade, and most trade barriers are reduced or eliminated outright. The World Trade Organization (WTO) is an agency of the United Nations which is responsible for supervising GATT agreements. While the WTO has generally been successful in promoting international agreements that cover trade between countries, its critics have nevertheless argued that GATT agreements reinforce the economic gaps between developed and less developed countries.

While GATT agreements cover most international trade today, in many cases trade between states is reduced or eliminated outright because of political considerations. Sometimes countries refuse to trade with one another because of political animosity. A boycott is a refusal to buy goods from a particular country, while an embargo is a refusal to sell them. For example, the United Nations imposed an embargo on trade with Iraq before and during the Persian Gulf War. The purpose of this embargo was to put pressure on the Iraqi government to comply with United Nations policy concerning inspection of weapons sites.

Since the GATT agreements were enacted, the question of whether most favored nation status should be granted to particular countries has occasionally been at issue. For example, in recent years, some have proposed that China be denied most favored nation status because of a consistent record of violating human rights. Likewise, opponents of South Africa's apartheid policy proposed to deny most favored nation status to that country. Today, such demands are no longer expressed because of South Africa's transition to multiracial democracy.

Another country that has been subject to trade embargoes and boycotts is Cuba. The Helms-Burton Act forbids American companies and in some cases, even foreign nationals, from doing business in or trading with Cuba, because of political opposition to Castro's government. This policy became controversial during debate over the implementation of the North American Free Trade Agreement in the early 1990s, because neither Canada nor Mexico had imposed such stringent sanctions.

4.4 Geopolitics and the less developed countries

As we saw in earlier chapters, formal geopolitical theory tended to pay little attention to the less developed countries. To be sure, the views of the various European powers concerning the role of the less developed countries were very different. The British, as we have seen, emphasized the role of the British Empire as a cornerstone of Rimland power against the Heartland. The Germans, on the other hand, viewed *lebensraum* as more important than the possession of overseas colonies. Yet neither view took into account the views of residents of the less developed countries themselves.

Once the less developed countries began to achieve independence, many political leaders and scholars within less developed countries began to address geopolitical questions from a Third World perspective. Of course, much of this investigation focused on questions unique to individual countries. Thus Nigerian analysts focused on Nigeria's role within Africa, for example; while those in India focused on India's relationships with Pakistan, Sri Lanka, China and other Asian countries.

At the same time, some within the less developed countries began to conceptualize the less developed countries as a group of states with geopolitical interests distinct from those of the developed countries of the world. Thus was born the philosophy of "North–South" geopolitics – a set of geopolitical views that stemmed from perceptions of the legacy of colonialism and its impact on the growing gap in the standard of living between the developed and less developed countries. The term "North–South" refers to the fact that most of the less developed countries are located geographically to the south of the majority of developed states.

4.4.1 THE LESS DEVELOPED COUNTRIES AND THE COLD WAR

In developing "North–South" geopolitics, scholars in the less developed countries contrasted their views with the geopolitics of the Cold War, which was in full swing when much of Africa and Asia was being granted political independence. The North–South perspective is often addressed in comparison with the "East–West" perspective, contrasting the capitalist and democratic countries of North America and Western Europe with the communist planned economies of the Soviet Union and her satellites.

The fact that much of the so-called Third World achieved political independence during the

BOX 4.3

The world grain trade

Trade in food is a little-noticed but often significant aspect of international trade, with important political implications. Over the twentieth century, there has been a rapid increase in both the production and trade in food and food products throughout the world. Prior to the early twentieth century, most of the world's countries were self-sufficient in food. Many less developed countries produced surpluses which were shipped to the developed countries. Western Europe, for example, was a net importer of food, while South America was a net exporter.

Since 1950, the global pattern of trade in agricultural commodities has changed dramatically. Latin America, Africa and Eastern Europe became net importers of food, while Western Europe became a net exporter of food. In general, the developed countries are now exporting food to the less developed countries. The colonial pattern of less developed countries exporting foodstuffs to the developed countries has been reversed.

Why has this reversal occurred? Although food supplies have increased, much of the increase in supply in less developed countries is offset by increased demand for food associated with rapid population growth. Moreover, in many countries, food production for export is more profitable than production for local consumption. Increasing numbers of farms and ranches in Costa Rica, Colombia, Brazil and other countries are now devoted to production of beef cattle for export to the United States and Europe.

International trade in agricultural products is influenced by political considerations. Countries encourage trade with friendly countries and discourage or prohibit trade with enemies. For example, during the Cold War the United Sates often used food as a form of foreign aid to try to encourage less developed countries to support American foreign policy objectives. Food as aid secures the disposal of agricultural surpluses in developed countries. It is regarded as a humanitarian gesture, a means of promoting economic development in less developed countries and a means of serving the short- and long-term diplomatic interests of the United States government.

Interestingly, the countries that received the most food aid were those critical to American foreign policy objectives at various points in time. During the 1950s, Europe and especially Greece and Turkey, which had long been recognized as critical to American strategy in the Cold War, received large shipments. In the 1960s and 1970s, Egypt, Israel, Yugoslavia and Poland were the leading recipients of food aid.

Cold War meant that the Cold War itself inevitably affected the development of geopolitics in less developed countries. The less developed countries reacted geopolitically to the Cold War in two fundamental ways. Recognizing that the United States and the Soviet Union were competing for influence outside the developed world, leaders in some of the less developed countries worked to play the two superpowers off against each other to their own advantage. Others rejected the philosophy of the Cold War entirely. The latter view led to the development of non-alignment, the development of formal organizations to advance the interests of the less developed countries, and eventually to the philosophy of the New International Economic Order.

Throughout the 1950s and 1960s, both the United States and the Soviet Union worked to secure alliances with newly independent less developed countries. In order to do so, both superpowers offered substantial amounts of economic assistance, military aid and support in local conflicts. Recognizing that both sides were offering assistance in return for their support, leaders of some less developed countries soon realized that it would be to their advantage to play the two sides off against each other. Countries that consistently supported either the United States or the Soviet Union found themselves taken for granted, while those that appeared to vacillate between support for the two superpowers found that both sides increased their offers of development and military assistance. Hence the history of many less developed countries is one of several years of friendship with the Soviet bloc followed by several years of friendship with the West, or vice versa. This point was brought home to American military personnel during the Persian Gulf War in 1991, when Iraqi soldiers who were captured by American troops were found to be in possession of American-made weapons, which Iraq had obtained during an earlier period of friendship between the two states.

Not all less developed countries were as eager to manipulate their positions relative to the Soviet Union and the United States, however. Instead, influential leaders of some less developed countries developed a philosophy that rejected the Cold War. Blaming the Cold War for nuclear proliferation and continued underdevelopment in the less developed countries, leaders such as Tito in Yugoslavia, Sukarno in Indonesia, Kwame Nkrumah in Ghana, Jawaharlal Nehru in India and Julius Nyerere in Tanzania developed a philosophy known as non-alignment.

The philosophy of non-alignment rejects the view that any less developed country must choose between support for the East and support for the West. Instead, the non-alignment philosophy suggests that the less developed countries should be recognized as an independent interest group on the global political stage, and that geopolitics should be more explicitly linked to efforts to redress historical and contemporary imbalances in economic status and standards of living between the developed and the less developed countries. According to the non-alignment philosophy, international politics cannot be addressed independently of the changing world economy. Leaders of the non-aligned nations "accused the West of maintaining the separation of politics and economics in international relations as a device to perpetuate its 'hegemony'" (Righter 1995: 93). Although non-alignment is not a formal political movement, leaders of non-aligned countries have met every few years to discuss solutions to common problems.

4.4.2 THE PHILOSOPHY OF NON-ALIGNMENT AND THE NEW INTERNATIONAL ECONOMIC ORDER

Several factors influenced the development of the philosophy of non-alignment. The major influences on the development of this philosophy include the legacy of colonialism along with the actual and potential effects of the Cold War on less developed countries. As we have already seen, most of the world's less developed countries went through a period of European colonization. After political independence was granted, the less developed countries remained far behind the colonial powers in economic strength. Many in the newly independent former colonies blamed their lack of development on the policies of the European powers, which had maintained colonial empires for their own profit. Thus a central component of the non-alignment philosophy was the view that the developed countries are ethically obligated to redress the legacy of the colonial past.

The Cold War itself also influenced the philosophy of non-alignment. Not only did proponents of non-alignment reject the geopolitics of the Cold War, but many blamed their dependency and lack of development on the Cold War. As we have seen already, many believed that Western Europe was a potential battleground between East and West. Yet these predictions did not come to pass, and despite the threat of nuclear holocaust Western Europe enjoyed several decades of peace and prosperity during the Cold War. In fact, the actual battleground of the Cold War was the "Third World".

Over the course of the Cold War, the less developed countries experienced numerous armed conflicts. Wars, insurrections and revolutions in the Middle East, Central America, Southeast Asia and elsewhere resulted in millions of military and civilian casualties, forced millions of other persons to leave their homelands as refugees and caused untold damage to property, buildings and the natural environment. These considerations led some to argue that the Cold War itself was partially responsible for the miseries of the less developed countries. By focusing the world's attention on the redress of economic imbalances between the colonial powers and their former colonies, proponents of non-alignment hoped to deflect geopolitical attention away from the Cold War.

4.4.3 NON-ALIGNMENT AND THE UNITED NATIONS

The philosophy of non-alignment became well established among many leaders in less developed countries during the Cold War. During this period, the United Nations emerged as a forum for promotion and in some cases implementation of the non-alignment philosophy. The United Nations had been established in 1945. During the early years of its history, one of its more important activities was to supervise the decolonization of European colonies. One of the more significant disputes that occurred during the San Francisco conference at which the United Nations Charter was written involved the status of the world's colonies. The European colonial powers blocked a proposal to place all of the world's colonies under the trusteeship of the United Nations (Meissner 1995). The Charter did recognize the right of colonies to self-government, and it authorized the General Assembly to manage decolonization.

Most former colonies joined the United Nations after achieving independence. The number of United Nations members increased from 51 in 1945 to 159 in 1985, with a large majority of new members located outside the developed world. The influence of these new members placed pressure on the European powers to grant independence to their remaining colonies. In 1960, the General Assembly adopted the Declaration on the Granting of Independence to Colonial Countries and Peoples. This declaration asserts that colonialism "constitutes a denial of fundamental human rights". National self-determination was thus recognized as a fundamental right, and the moral authority of the United Nations was placed behind independence movements around the world. The Declaration was invoked in support of independence struggles in several parts of the world. In Africa, for example, the United Nations took an active role in promoting the independence of colonies such as Angola, Mozambique, Zimbabwe and Namibia despite the resistance of the colonial powers.

Economic development has been a focus of the United Nations throughout its history. Many of the economic activities of the United Nations have been focused on reforming the global economy in order to redress imbalances in standards of living between developed and less developed countries. In 1964, the General Assembly established the United Nations Conference on Trade and Development (UNCTAD). The purpose of UNCTAD was to promote world trade and to coordinate trade and development policy. Shortly before the first session of UNCTAD was convened, 77 less developed countries signed a statement calling for UNCTAD to address efforts to promote development in less developed states. These 77 countries became known as the Group of 77, and in effect the Group of 77 became an interest group promoting the non-alignment philosophy within the United Nations.

The positions on international economic development espoused by the Group of 77 eventually became codified into what is now known as the New International Economic Order. The principles of the New International Economic Order were written into a series of resolutions passed by the General Assembly in 1974 and 1975. The goals of the New International Economic Order include greater economic self-sufficiency for less developed countries, increased participation in trade and industrialization, regulation of the activities of transnational corporations, and protection of the resources of less developed countries. In order to achieve these goals, representatives of non-aligned countries proposed to stabilize commodity prices, eliminate protective tariff, and fix terms of trade so that the developed countries could not profit from the added value associated with exchanging manufactured goods and finished products for raw materials. Such policies were justified by representatives of the less developed countries on the grounds that they "were owed as compensation for the accumulated Western privilege and prosperity built on tropical sweat" (Righter 1995: 104).

The principles of the New International Economic Order remain controversial, however. Indeed, the formal declaration of the New International Economic Order in 1974 "essentially dismisses the entire framework of international law and practice as an 'unequal treaty', a device whereby the wealthy impose their liberal philosophies and free-market doctrines of competition on the socially vulnerable and economically uncompetitive" (Righter 1995: 110). Not surprisingly, such arguments were resisted in many of the developed countries. Those opposing the concept of the New International Economic Order came to regard the United Nations as an instrument of the developing countries rather than a meaningful forum for the resolution of international disputes.

This criticism extended to the activities of other United Nations agencies as well. For example, the United Nations Educational, Scientific and Cultural Organization (UNESCO) was established, according to its charter, in order to "contribute to peace and security by promoting collaboration among the nations through education, science and culture in order to further universal respect for justice". Among the activities of UNESCO are the promotion of literacy and education, and the exchange of scientific information and mass communications. Yet some criticized UNESCO's activities as biased in favor of the less developed countries. The United States and Britain withdrew from UNESCO in the mid-1980s.

4.5 Geopolitics and the less developed countries after the Cold War

As we have seen, the less developed world's views of geopolitics were conditioned by the Cold War. Now that the Cold War has ended, will the geopolitical position of the less developed countries change? How might such changes affect geopolitical thinking in the years ahead? In order to address these questions, it might be useful to consider how the original rationales underlying the geopolitics of the less developed countries have been affected by the end of the Cold War. The ending of East–West hostilities changed the geopolitical position of less developed countries dramatically. The arguments that the Cold War contributed to underdevelopment were now moot. It was no longer argued that the Cold War was contributing to violence and warfare in the less developed countries. Yet less developed countries could no longer rely on playing the superpowers off against one another. Instead, the increased dominance of the world economy by the capitalist democracies sometimes reduced development prospects.

Within any capitalist society, capital flows into those regions and economic sectors whose profit potential is highest. The collapse of communism in the former Soviet Union and Eastern Europe has opened numerous new markets to Western investment. Because these countries are located relatively close to Western Europe and North America and contain literate and well-educated but poorly paid populations, they were soon recognized by Western-based multinational corporations as lucrative targets for foreign investment. Moreover, the developed countries no longer have the incentive to "buy" the friendship of the less developed countries. Accordingly, it may be that investment capital that might at one time have been sent to less developed countries is instead being sent to better-developed former communist states in Russia, Eastern Europe and elsewhere.

Plate 4.3 A traditional food stand in Macao (photo: Kathleen Braden

If we are to predict whether the geopolitical position of the "Third World" is to improve in the years ahead, we must address whether we can expect economic development prospects in the less developed countries to improve now that the Cold War has ended. Development prospects could improve in either of two ways. First, can the principles of the New International Economic Order become a reality in international geopolitics? Second, can the less developed countries develop independently of the New International Economic Order?

5

Nationalism and Culture

KEYWORDS

nationality,
ethnicity,
multinational state,
supranationalism,
culture,
irredentism,
enemy,
minority

KEY PROPOSITIONS

- Two apparently opposing trends are evident with respect to national identity: both re-emergent nationalism and new supranational identities are emerging on the globe.
- The concept of "enemy" is often a useful centripetal force for a nation.
- Most states in the world today contain minority groups; provision of rights to minorities is key to avoiding internal conflict.
- The idea of a nation-state may be increasingly an artifact of the past on the globe.

5.1 Introduction

As noted in Chapter 2, there are approximately 200 states in the world today, but they contain persons who belong to thousands of distinct nations. The precise number of nations, however, depends on definition. Should nationality be determined by language, culture, religion, or historic allegiance? During the Soviet period, American students living in the USSR were often asked the question: "What is your nationality?" but the reply, "American" was insufficient. Soviets would look puzzled and say, "Yes, we know you are Americans, but what is your nationality?" Further conversation revealed that what Americans refer to as ethnic heritage, Soviets thought of as nationality.

Nationality is a rather vague expression of group identity, perhaps at the largest level after family, clan, or tribe. The concept of nationality was born out of the European nation-state system, which emerged in the late Middle Ages. Yet in other parts of the globe, identity of group, rooted in part on location, has also been evident in tribal areas or units as large as empires. Great "nations" of people as far back as Biblical times have made their appearances and disappearances on the world map. For example, the Khazars were a nation of Turkic people whose empire lasted several hundred years in Eurasia; yet today, no one identifies him- or herself as Khazarian, and the language and culture have disappeared from the world.

While the idea of "nations" has existed for millennia, the modern concept of "nationalism" only dates back several hundred years, and may already be outmoded. Nationalism has become a troubling concept in the post Cold War world. Division of much of the globe into easily identified blocks of East versus West made geopolitics a more clear-cut discipline than a more chaotic

Plate 5.1 Women in Samarkand, Uzbekistan (photo: Kathleen Braden

Plate 5.2 Tajik family in Uzbekistan (photo: Kathleen Braden)

world wherein nationalist fervor has been reborn in many regions. Harm De Blij (1992) has noted: "the world of the twenty-first century is burdened with a boundary framework that is rooted in the 1800s, and with political systems few of which can accommodate the stresses of resurgent nationalisms and regionalisms".

Two counter trends seem to be emerging at the end of the Cold War. On one hand, national identities that had often been submerged in constructed countries (such as the USSR and Yugoslavia) are now calling out again in force and resulting in conflict areas on the world stage. On the other hand, supranationalism, or the tendency to form alliances and political groups larger than national identities, continues and perhaps strengthens. This latter process may be the end of a trend that has been occurring in some global areas for many centuries. Westphalian identity has given way to German, which may be giving way to "European". But some suggest that supranationalism exists mainly in political or economic terms, not psychological definition. Symbols of national identity, such as language, flags or food, could not be superseded by pan-European symbols, despite the selection of a flag and anthem for the Union.

As ideas on national identity become blurred, so too do notions of competition and enemy-making. In the Cold War era, enemies were simple to discern. Nationality was of less importance than ideology or political alliance. Now, the clar-

BOX 5.1

Changing one's nationality without ever moving

The Reverend Zsigmond Gsukas is 75 years old and lives in the village of Samorin in the country of Slovakia. Although this elderly pastor has never lived outside the region, on January 25, 1993, *The New York Times* reported that Mr Gsukas has been a citizen of five different states during his lifetime. He was born as a citizen of the Austro-Hungarian empire, became a citizen of Czechoslovakia after World War I, then of Hungary after 1938, then Czechoslovakia again, and in 1993, a resident of Slovakia. His identity as a Hungarian has not changed, like the nationality of almost five million Hungarians in Central Europe living outside the legal boundaries of the state of Hungary.

BOX 5.2

The Khazars: a nation lost in time

From the seventh to the tenth centuries AD, the Khazar Khans (rulers) controlled a vast area from the Caspian Sea all the way to the Dnieper and Oka Rivers. The Khazar nation spoke a Turkic language, but had converted to Judaism in the ninth century. They benefited from all the trade between Asia and Europe which traversed their territory, forcing traders to pay heavy taxes. In the tenth century, the Kievan king, Sviatoslav, who was in the process of building a new great nation of people out of various Slavic tribes, defeated the Khazar Khan.

BOX 5.3

Money as the symbol of a nation

With the advent of paper money, national pride is often at stake when the question arises about what image should be portrayed on currency. By 1999, the European Union countries were due to begin the switch to a single bill, at first called the European Currency Unit (or ECU), but some Germans complain that *"kuh"* means "cow" in German and another title should be picked; in 1996 the term "euro" was selected. What famous Europeans should be pictured on the ECU? Because of nationalism among member states, the European Union finally decided on an abstract symbol, rather than a person's portrait. Independent Kyrgyzstan traded in a portrait of Lenin on the ruble for one of the hero of their national epic, "Manas" on the *som* note; and even some British citizens complained that national pride was injured when new currencies updated (and therefore aged) the traditional portrait of Queen Elizabeth on the pound.

ity of that era is being replaced by a period of more complicated nationalisms and conflicts.

5.2 What is nationalism?

As we noted in Chapter 2, nations are defined on the basis of culture, language, ethnicity, and religion. The word "nation" derives from the Latin root indicating "to be born" and initially referred to a stock or breed of people.

As the modern state system emerged in Europe in the late 1600s, a national consciousness also developed. Initially, one's allegiance was probably to a local family group or region, and later, to a feudal lord or king. Sovereignty was truly vested in the sovereign, or ruler. The best definition of "nation" may be that it is a group of people with a common heritage or a common culture. Glassner and De Blij (1989) refer to this as a group sharing one or more important culture traits like religion, language, political institutions, values, or historical identity. In his discussion of a "Fourth World", where thousands of nations on earth have been subdued, often violently, to forced participation in the boundaries of a legal state system, Bernard Nietschmann (1994) notes, "The term nation refers to the geographically bounded territory of a common people as well as to the people themselves. A nation is a cultural territory made up of communities of individuals who see themselves as 'one people' on the basis of common ancestry, history, society, institutions, ideology, language, territory, and often religion." Fourth World theory suggests that boundaries of nations are more long-lasting and meaningful for people's lives than state borders, and that history has shown a tendency for the modern state system to suppress what would otherwise be the natural boundaries of nations. Journalist Joel Garreau in his (1992) book, *The Nine Nations of North America*, likewise argued that the real borders of nations on the continent are dependent on cultural ties rather than political boundaries.

The concept of nationhood is often linked with the places in which cultures associated with nations arise. Over the past several centuries, large-scale movements of peoples have affected the meaning of nationhood. The idea of nation may be difficult to maintain outside of set spatial boundaries constituting a homeland, as cultures meet and mix and change, and nationalities in turn are modified. For example, would Irish-Americans consider themselves more "Irish" or more "American" by national identity? Major nations exist even in a scattered state among many political boundaries (Kurds, Jews, Han Chinese, Uighurs). There are very few examples of pure nation-states in the world today. Much more common on the world map is the multinational state, the mixture of nationalities into one political unit which may or may not be cohesive over time. The presence of various nationalities within a state can constitute an important centrifugal force, tearing at national unity if sufficient attention is not paid to minority rights. Even states named for a nationality group may exist more in title than reality. Andorrans are a minority within the political state of Andorra, just as Buryats were a lesser group in the Buryat ASSR (Autonomous Soviet Socialist Republic) of the former USSR, outnumbered by Russians in their own titular region (see Table 5.1).

Many people living in multinational states have expressed a desire to attain greater political autonomy or independence. The expression of these goals is known as nationalism. A closely related concept is ethnicity. Ethnic groups are people who feel bound by a common culture and heritage, although their ties may also be associated with social perceptions of race.

Ethnicity is associated with territoriality in that spatial identity may be an important component of ethnic identity. The word "ethnicity" is from the Greek meaning "people" or "nation", and is also closely aligned with cultural traits such as language. Algerians in France may feel that ethnically they are Arab, but by citizenship, French. If they have been in France long enough to develop a sense of national identity, they may also regard themselves as French by nationality. People in Bulgaria who are ethnically Moslems, but Bulgarians by nationality, may feel distinct from other Bulgarians because of religious conversions that occurred generations earlier, even though by other measures they are indistinct from their fellow Bulgarians. Within the modern state of Bulgaria, there are 200,000 Bulgarians who are Moslem and one million Bulgarian citizens who are also Moslem but Turkish by nationality.

Another example of confused identities over ethnic and nationality heritage occurs in the state of Somalia. In the past, this country was often cited as one of the few examples in Africa of a nation-state, a country composed ethnically of only Somalis, but apparently the meaning of the term was more clear to outsiders than to the Somalis, who continued to distinguish themselves largely by clan identity, despite a common language and Islamic faith. Somalia deteriorated into open warfare between major clans in the 1990s as the world watched the mounting casualties and disintegration of a central government.

As noted in Chapter 2, a lack of correspon-

Table 5.1 Multinational life within the Buryat ASSR of the USSR, 1989

Nationality	*Numbers*	*Nationality*	*Numbers*
Russian	726,165	Kyrgyz	208
Buryat	249,525	Lazgini	206
Ukrainian	22,868	Chinese	191
Tatar	10,496	Ossetian	186
Belorussian	5,338	Eveni	182
Armenian	2,269	Khakasi	176
Azeri	1,679	Bulgars	133
Evenki	1,655	Avartsi	129
Chuvash	1,307	Chechen	126
Mordvinian	1,294	Greek	124
Kazakh	1,270	Estonian	117
Jews	1,181	Kalmyk	106
Uzbek	994	Turkmenian	75
Bashkir	920	Dargintsi	69
Moldavian	912	Komi	69
Yakut	705	Ingush	59
Georgian	612	Komi-permyaki	55
Lithuanian	604	Kabardintsi	55
Udmurt	524		
Tuvinian	476		
Marii	388		
Gypsies	250		
Latvian	239		
Tajik	210		

Source: Goskomstat SSSR pp. 11–12, Vsesoyuznaya perepis' naseleniya 1989.

BOX 5.4

Two major nations without independent states

The Kurds make up one of the largest nations in the world not to have its own state. More than 20 million Kurds live in Turkey, Iran, Iraq, Syria, Armenia, and Azerbaijan. Their language is distinct within the region, but classified within the "Iranian" group. For a brief few months after World War II, a Kurdish republic existed in northern Iran, but a series of conflicts with Turkey, Iran, and Iraq have prevented the realization of a "Kurdistan" state. Today, there is even fracturing within the Kurds themselves.

In a similar position are the Uighur people, a Turkic minority group in western China numbering almost 10 million adherents to the Moslem faith. The nation has a history dating back more than 1,000 years and includes a substantial minority population across the border in Kazkahstan. In the 1990s, unrest broke out in Xinjiang Province as Uighurs called for the establishment of a greater "Uighurstan", a notion not much in favor with the Chinese or Kazakh governments.

dence between state boundaries and areas of national dominance has resulted in many international conflicts. Nationalism may create trouble on an interstate level when a group presses irredentist claims on another territory. Irredentism is the desire to bring into a state all areas that had once been part of it or areas where members of the nationality group live. Originally, the term was used to refer to an area in northern Italy which remained part of Austria in 1871. Italian nationalists referred to the region as *Italia irredenta* ("unredeemed Italy").

Areas of the world today which may be subject to conflict due to irredentist sentiments include Central Europe, where Hungarians inhabit states neighboring Hungary in significant numbers; northern Kazakhstan, where the Russian population outnumbers Kazakhs and has agitated to be joined with Russian Siberia; northwest Pakistan, where the Pathan people are linked by nationality to people in Afghanistan; and the Caucasus, where Armenia and Azerbaijan have both expressed irredentist claims on exclaves of the other state (see Figures 5.1–5.3).

Irredentism has proven to be a serious source of wars, as it often serves as a pretext for invasion. Iraq's 1990 invasion of Kuwait was in part based on unsatisfied irredentist claims to Kuwait territory dating back to the time of British control.

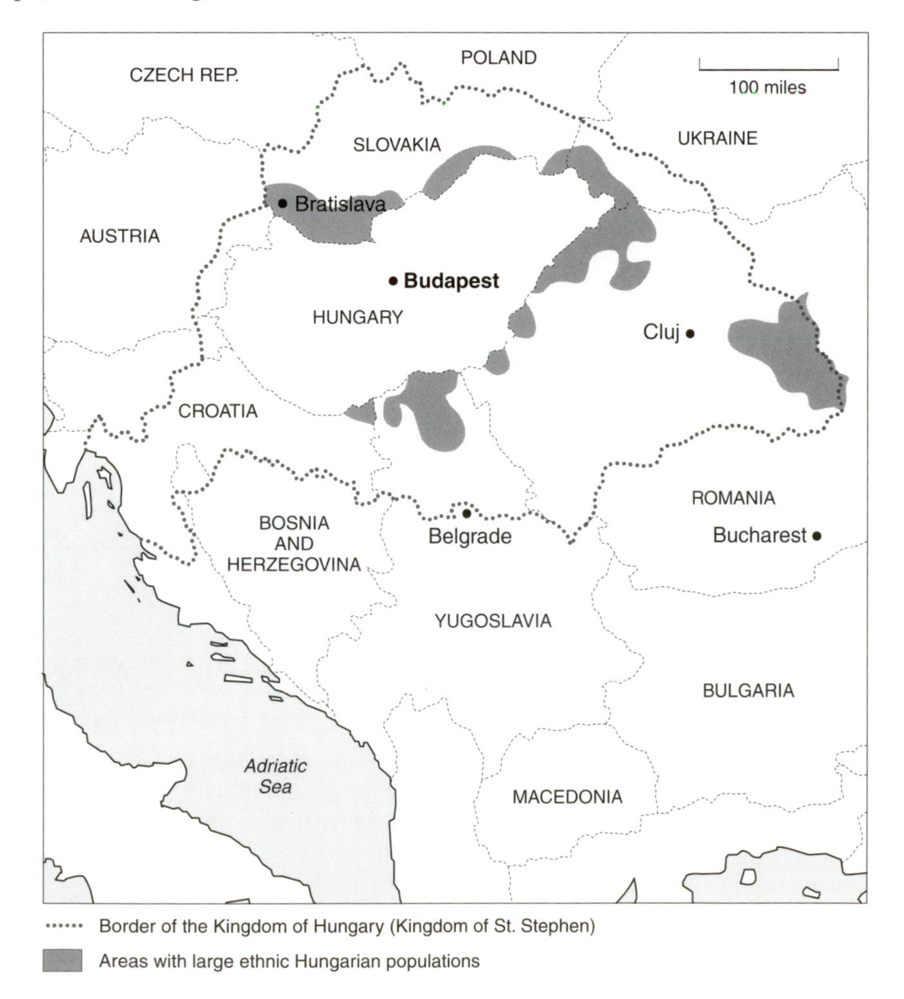

Figure 5.1 Irredentist potential. The largest single minority group in Europe are ethnic Hungarians who do not live within the boundaries of present-day Hungary. Of the 15 million Hungarians in Europe, less than 11 million live in Hungary. They form 7 per cent of the population of Romania, and 11 per cent of the population in Slovakia

Figure 5.2 North Kazakhstan is mainly populated with ethnic Russians, many of whom agitate to become part of Russia

Figure 5.3 Armenia and Azerbaijan: conflicting claims. Nagorno Karabakh is an enclave of Armenia within Azerbaijan; Nakhichevan is an enclave of Azerbaijan within Armenia

5.3 Nationalism and perception

After the devolution of the USSR, many ethnic Russians are contemplating (often in depression) what the Russian national consciousness is all about. A Russian friend wrote from Siberia:

> I for some reason fear that Russia has lived out its century, that her sun is setting. Put us at a table with a glass of vodka with Kazakhs, Tatars, Yakuts and within half an hour we will be embracing. But with the Baltics, the West – this is a suitcoat with strange shoulders. Alien also, but all the same closer – is the East. But not Buddha, Krishna, or Chinese hieroglyphs. No, Russia has its own sorceress. It is the soul and body of Russia, her choice and her cross. And we are all the poorer for the fact that we do not understand this. We are always measuring ourselves by the western coat or the Chinese hieroglyph. Russia has, as every person has, its own unique face, but we are always putting on make-up to try to look like someone else.

Plate 5.3 Multilingual (Russian, Georgian and Abkhazian) sign in the former USSR (photo: Kathleen Braden)

Plate 5.4 Multilingual (Russian, Georgian and Abkhazian) sign at the Sukhumi Botanical Garden, Georgia (photo: Kathleen Braden)

BOX 5.5

Did a Soviet nationality exist?

Hundreds of different nationality groups existed in the USSR, all recognized as separate nations by the central government, which conducted censuses to determine nationality and insisted on placing national identity on internal passports. With the devolution of the USSR into 15 republics in 1991, titular nationalities once again reclaimed spatial units, but this did not mean that inter-republic migration settled out into nation-states. The scrambled eggs that made up the Soviet empire could not be so easily unscrambled, and even after the creation of a "Georgia" or an "Estonia", nationality and citizenship problems remained to be solved. Russian nationals in the Baltics try to learn Latvian or Lithuanian languages; Russians in northern Kazakhstan agitate to be joined with Siberia. Meanwhile, people who truly considered themselves "Soviet" in nationality through intermarriage have been set adrift in a sea of devolution. Citizens of Kazakhstan whose fathers are Kazakh and mothers Russian must now choose which nationality will be on their passports and thus set their own fates into the hard legal language of citizenship.

As we have seen, most modern states are multinational. Nationalism and ethnic conflict are centrifugal forces that can tear a country apart if unresolved. In response, states often create centripetal forces intended to promote unity, with such symbols as national anthems, mottoes, flags, Olympic teams, or loyalty oaths. Symbolic expressions of nationalism on the landscape often take the form of monuments, historic spots, or place-names. Geographer Wilbur Zelinsky has suggested that place-names with patriotic significance are especially important in the early stages of nation-building (Zelinsky 1989). An example of this phenomenon may be found in Israel, where Hebrew (Jewish) place-names, such as Yerushalayim for Jerusalem (as opposed to the Arabic Al-Quds), are important symbols in the Jewish–Arab struggle for actual and emotional control over territory. Likewise, states of the former USSR are experiencing with independence a rush of renaming places in local languages or reviving pre-Soviet names. The city of St Petersburg, for example, had been renamed Leningrad by the Soviets in honor of Lenin, the architect of the Russian Bolshevik Revolution, but after the collapse of the Soviet system, the city name was restored.

As noted in Table 5.2, national groups may have a different name identity for their state than the one commonly translated into another language. In Finnish, the word "*Suomi*" refers to lakes and is more descriptive of the homeland than the "land of the Finns" term which became the outsider's name for the state.

Words are important symbols of national identity because language is one of the more potent elements that define nationality. Culture is not biologically determined, although racial characteristics may be perceived as an element of culture; rather, it is learned behavior. Expressed in terms of tangible items or artifacts, such as artwork, dress, or traditional foods, culture may also be a social or mental construct (language, games, or religion). Perhaps the broadest definition may be that culture is an entire way of life of a people.

An example is the development of the French nation, which went hand-in-hand with the emergence of French culture, especially language and literature. The Franks, originally a Germanic tribe, settled in the region of the Roman Empire called Gaul in the fifth century. The term for their nation, "Franks", meant "freemen", and they were divided into two groups: Salians along the North Sea, and Ripuarians on the banks of the Rhine. As early rulers such as Clovis and Charlemagne carved out notions of empires, the French language began to develop out of its Latin roots. The "Serments de Strasbourg" oath taken by successors to Charlemagne in AD 842 is considered one of the first written texts to represent a separate French language.

Even today, French nationalists fight to preserve their heritage on the battlefield of words. "Franglais", or the invasion of English phrases into the French language, is actively discouraged by members of L'Académie Française, which has set standards for the French language for nearly 300 years. Beginning in the early 1970s, the

Table 5.2 Nationalism and state names: Europe

State name in English	*Meaning*	*Name in native tongue (and meaning)*
Finland	Swedish: "Land of the Finns"; maybe from German *finden* ("to seek")	Suomi (lake,swamp)
Georgia	Maybe from St George	Sakartvelo
Armenia	From Armenak legend	Hanrapetut'yun
Austria	Latin version of "eastern kingdom"	Osterreich (eastern kingdom)
Croatia	Maybe from *sharvatas* ("armed")	Hrvatska
Greece	*Gra* ("venerable")	Elliniki
Hungary	Ugrian people	Magyar (landsmen or natives)

Sources: Britannica yearbook 1993; Adrian Room, *Place-Names of the World*, Angus & Robertson publ, London, 1987.

Dictionnaire des Termes Officiels began to list French terms that should be used by government officials to set an example to others in the preservation of the language. (Examples recently added include *essaimage* for spin-off, *disque compact* for CD, and *balladeur* for Walkman).

Not all struggles for dominance between nationalities may take place in the realm of territoriality. As borders become more and more easily transcended in the modern world, cultural invasions may be almost as potent a method for extending national power and influence. English is considered the most widely spoken language on the planet today, despite the term *lingua franca*, or "common language" left over from earlier centuries when French was the language of colonial commerce.

Author John Rockwell pointed out the importance of the role of American culture throughout the world today, and concluded:

> American popular culture has never been more dominant internationally nor more controversial. American cultural exports swell at least some of us with pride and help reduce the trade deficit. Serious money is involved; this is America's second-biggest export after aircraft ... Maybe Western suits worn by Saudi or African businessmen, maybe even the English language itself, are not so much emblems of American superiority as the simple acceptance by a developing world of a single international standard of discourse.
>
> (*The New York Times*, January 30 , 1994)

If power implies not only military might or control over territory, but the ability to influence the behavior of another state's government or another nation, then American hegemony over popular culture in much of the world may imply, for better or worse, a new extension of national might. Samuel Huntington extended this idea in his 1993 *Foreign Affairs* article, arguing that future conflicts on the globe may be wars of culture, rather than between states.

The irony of the geopolitical world at the end of the twentieth century may be seen in the fact that nationalism is still an important force, yet single-nation states are increasingly rare; nationalist sentiments swell even as supranationalist organizations exert more influence over daily lives; and calls for preservation and celebration of nationality arise despite increasing conformity and convergence of cultures.

5.4 The concept of "enemy"

Whether or not centripetal forces are linked to nationalist considerations, their purpose is generally to establish a common identity. At times, this process is served by the creation of a real or imagined enemy, generating a sense of "us" versus "them", the idea of otherness discussed in Chapter 1.

In all cases of international conflict that become violent, the notion of "enemy" is vital. It is the psychological counterpart of a fence that divides territory and creates a notion of insider and outsider, friend and stranger, them and us. Psychologist Sam Keen wrote that the idea of "enemy" is society's most potent weapon: "In the beginning we create the enemy. Before the weapon comes the image. We think others to death and then invent the battle-axe or the ballistic missiles with which to actually kill them" (Keen 1986).

The ethical nature of enemy notion may be open to question, but the utility of the concept in nation-building and state power seems clear. Having enemies is a useful way of defining nationhood and national purpose. Various elements of culture, particularly in oratory and art, help contribute to creating an enemy image which can then be reinforced by unfolding events.

Frederick Hartmann has attempted to explore the notion of enemy as a concept that has utility in understanding a system of international relations and conflict. He argues that there is a tendency toward "conservation" of enemies in geopolitical relations: one state should not take on more enemies than it can deal with practically at any given time. Few states escape the idea of having any enemies because of the utility they provide in creating national cohesion. He writes:

> In a rare case it is possible to imagine a nation with no active fear of being attacked, but in practice there is always someone. This focusing of concern on particular potential enemies is a first characteristic of the system and of the international environment to which states must relate. Such behavior is "natural" to the system,

> then, although each nation's list of enemies varies ... By such a list, a nation defines its power problem: those foreign nations considered a source of potential attack.
>
> (Hartmann 1982: 65–66)

What are the sources of enemies in geopolitics? Hartmann argues that there are many cases of nations or states which seem to have experienced a disproportionate number of enemies or conflicts compared with the planetary norm, and looks for explanation. The first factor he considers is geographic: the location of states. Conflict over borders and territory is common enough to suggest that the prime candidate for "enemy" might be "neighbor", although distortions to the distance factor in this notion are created due to colonial expansion and overseas extension of military power.

Does the utility of enemy notions mean that real enemies do not exist? There are certainly many unfortunate examples in human history of attempted genocide, or systematic elimination of a national group. But it is important to note here that whether threats are quite real or merely perceived, once the notion of enemy is established, it becomes a tangible element in geopolitics. Some enmities have persisted for centuries, such as conflict in the Balkan peninsula between Serbs, Croatians, and Moslems. In other cases, public perception may give way quite quickly, as in the case of the American perception of Russians at the end of the Cold War. In part, the durability of an enemy group depends on maintenance of threat, the cycle of revenge, and public culture. The United States has had a series of "enemies" throughout its history in its struggles for power and territorial control, although loss of life was perhaps most brutal during the American Civil War. As the enemies of World War II, Japan and Germany, gave way to the ideological enemy of the communist during the Cold War, the United States found its identity as a nation in part determined by its opposition to the Soviet system for 40 years. Proxy enemies in Korea, Vietnam, Iran, and Grenada appeared during this period, and participation in military responses was always justified on the basis of the overarching Cold War face-off. The Russian government was often portrayed in political cartoons of the period as a drooling, hungry bear, ready to take over much of the globe. Soviet newspaper cartoons in turn portrayed Uncle Sam as a power-crazed and greedy capitalist with dollar signs for eyes in contrast (see Figure 5.4 for an Arab view of American political leaders).

Figure 5.4 Creating enemy image: a portrayal of the control of US presidential candidates by Israel during the 1988 election

In a 1949 prophetic article, later reprinted by *The Atlantic Monthly*, Archibald MacLeish noted the danger in defining US purpose as a nation solely in terms of what we were not:

> by putting hatred and fear of Russia first we have opened the sacred center of our lives, our most essential freedoms – the freedom of mind and thought – to those among us who have always hated those freedoms ... A people who have been real to themselves because they were for something, cannot continue to be real to themselves when they find they were merely against something.
>
> (*The Atlantic Monthly* March 1980: 40)

With the loss of a clear, powerful, and well-defined enemy state at the end of the Cold War, Americans had to make do with enemies who were smaller in stature: Noriega in Panama, Castro in Cuba, Kim Il Sung in North Korea, Hussein in Iraq, Gaddafi in Libya, and finally, even clan leaders in Somalia.

Racism is also a powerful reinforcing factor in creating a successful enemy image. By the 1980s and 1990s, some Americans were expressing opinions on Japan as the chief "rival" of the United States in world affairs; helped perhaps by relatively recent memories of World War II. Most people expressed this concern in economic terms, but popular culture in media imagery helped to fortify the fears. For example, on May 9, 1988, *US News & World Report* featured a cover report entitled "Japan Moves In" and subtitled "Tokyo Inc's Bold Thrust into the American Heartland", and illustrated by a cartoon image of an industrial worker planting Japan's flag into the middle of Tennessee (due to the Nissan factory located there) and Japanese flags all around the state. The article on the inside was titled "A New Land of the Rising Sun" (see Figure 5.5).

Figure 5.5 A modern illustration of nationalism fears: the cover of *US News & World Report*, May 9, 1988 © US News & World Report. Reprinted by kind permission

If the notion of enemy is useful in international relations both for nation-building and perhaps preparation for actual conflict, then how is that notion created and reinforced in the population as a centripetal force? In recent years, studies by various social scientists and journalists, as well as a new disciplinary sub-group (political psychologists) have appeared to explore the relationship between individual human behavior and the behavior of states or nations with respect to enemy-creation. In the 1970s, the International Society of Political Psychology was founded and now numbers over 1,500 members.

A leading proponent of this approach to examining international conflict has been Vamik Volkan. His 1988 book, *The Need to Have Enemies and Allies*, examines the basic human need to form identity, and how the notions of outsider and enemy contribute to that development. Volkan then examines the move away from instinct to a more civilized form of national behavior, asking whether nations are capable of making that move. He recognizes the importance of enemy concepts to nationalism, and writes that patriotism is a learned, not an innate, human sentiment and therefore psychology should not be overlooked in trying to come to grips with a definition of nationalism. Citing a 1945 essay by George Orwell, Volkan (1988) notes that there are three important characteristics of nationalism, all related to the "enemy" or the "outsider":

1 obsession: even a small slur by the outside group is taken with utmost seriousness in the drive to reinforce national group identity;
2 instability: enemy selection is a fickle process, with many changes in partners (compare Hartmann's notion that enemies must be "conserved");
3 indifference to reality: atrocities can only be committed by "them", certainly not by "us"!

Political cartoons and posters from a variety of cultures and historic periods show that the creation of an "enemy" for a nation is a necessary prelude to killing another human-being during times of war, because our natural response may otherwise be refusal to kill. The pattern, therefore, that all cultures showed was an attempt to dehumanize the enemy – make him or her not only outside the national group identity, but also outside general humanity and therefore deserving to be eliminated. The dehumanization process took the form of portraying the enemy as animal, barbarian, criminal, enemy of God, a threat to culture, or merely an abstraction.

Unfortunately, the creation and portrayal of enemies has often resulted in discrimination against individuals whose cultural backgrounds or physical appearance are consistent with the stereotype of the enemy. Many German-Americans anglicized their names during World War I, and half the states enacted laws prohibiting the teaching of the German language in schools. Thousands of Japanese-Americans were interned during Word War II. After the American embassy was seized in Tehran in 1979, Iranians in the United States were vilified and subject to abuse. Author Fred Shelley recalls two highly respected professors, both American citizens, who were abused while riding public buses because strangers believed they might be Iranian (one was of Indian Brahmin ancestry and the other of Greek heritage).

During the Korean and Vietnam conflicts, Americans needed to believe that through the domino notion, there was a direct threat to the United States itself. But the difficulty in identifying the enemy in the Vietnam war, perhaps reinforced by issues related to racial differences, made for trauma among American troops, and often, tragic victims of Vietnamese civilians. The most infamous case was the My Lai attack in 1968. The target was a group of Vietcong insurgents, but the resulting action of the second platoon under the leadership of Lieutenant William Calley led to the deaths of up to 500 villagers, including women and children, who were killed outright by American soldiers. In 1969, a court martial convicted Lieutenant Calley of the murders of 109 Vietnamese. His conviction was later overturned by a federal district judge. At his court martial, Calley tried to explain the psychological atmosphere that influenced him:

> They didn't give it a race, they didn't give it a sex, they didn't give it an age. They never let me believe it was just a philosophy in a man's mind. That was my enemy out there and when it came between me and that enemy, I had to value the lives of my troops.

5.5 Minority groups

The concept of "minorities" in national populations suggests a mathematical measurement that may be illusory. Minority status implies not only lesser numbers, but also the perception that one is not fully part of the whole nation. If the idea of "majority" and "minority" were applied at a global level, the majority group in the world population would probably be best represented by young Asian females who are farmers by occupation. The word "minority" in the United States has often been taken to mean "non-Caucasian", but becomes less specific when, for example, Hispanic-Americans are considered a minority. The popular connotation of minority will surely be challenged in the middle of the next century when (if demographic statistics continue along the present trend) the average American will not be from the traditional European heritage groups considered to make up the "majority" in the United States.

The United Nations Human Rights Sub-Commission defines a minority as a group numerically smaller than the rest of the population of a state and in a nondominant position, whose members are distinct from others in the population and have a sense of solidarity for preserving their culture. Article 15 of the United Nations Universal Declaration of Human Rights states that every human being has a right to a nationality; Article 7 forbids discrimination; and Article 21 states that everyone has the right to take part in the government of his or her country.

Very few examples of true nation-states exist now on earth, as seen in Table 5.3 (in which the majority group is defined based on ethnic or linguistic bases). Therefore, every country has some minority population, including Japan and Iceland, which are often cited as the best examples of nation-states. In some African and Asian states, mainly as a result of colonial boundary policies, it is very difficult to determine which group constitutes a "majority" within the state's population

Table 5.3 Minorities in selected states

Name	*Majority*	*Minorities*
Algeria	Arab (82.6%)	4.6 million
Andorra	Spanish (50%)	28,550
Angola	Ovimbundu (37.2%)	6.6 million
Argentina	European (85%)	5 million
Australia	European (95.2%)	0.8 million
Bahrain	Bahraini Arab (68%)	0.17 million
Bangladesh	Bengali (97.7%)	2.53 million
Benin	Fon (39%)	3 million
Bosnia-Hercegovina	Bosnian (39.5%)	2.7 million
Brunei	Malay (68.7%)	83,800
China	Han (93%)	80 million
Finland	Finns (93.5%)	327,000
France	French (93.2%)	3.9 million
Greece	Greek (95.5%)	0.46 million
Guyana	East Indian (51.4%)	0.36 million
Iceland	Icelandic (93.9%)	16 000
India	Hindi (38.8%)	544 million
Israel	Jews (83%)	0.8 million
Italy	Italians (98.8%)	0.7 million
Ivory Coast	Baule (23%)	9.5 million
Jamaica	African (75%)	0.6 million
Japan	Japanese (99%)	1.2 million
Kyrgyzstan	Kyrgyz (52.4%)	2.2 million
Malaysia	Malay (61%)	7.2 million
Mozambique	Makua (47.3%)	7.8 million
Niger	Hausa (52.8%)	3.9 million
South Korea	Korean (99.9%)	44,000

Source: Encyclopedia Britannica (1993).

(see the Ivory Coast example in Table 5.3). On the other hand, the population of China is so large, that although only 7 per cent is not Han Chinese, the resulting number of "minorities" is equal to at least 80 million people.

What determines when a minority "nation" group seeks to create its own state through an insurgency movement or political agitation? Why do some minority groups, such as the Swedish-speaking population of Finland, seem content to remain Finnish citizens, while in the case of the Basque population of Spain, the world has seen a serious and often violent movement to secede and create a nation-state?

Marvin Mikesell and Alexander Murphy have provided a conceptual scheme to classify aspirations of minority groups and raise questions about measuring the effectiveness of state government policy toward treatment of minorities (Mikesell and Murphy 1991). While admitting that much discord in the world may be due to an inadequate number of existing states compared to nations, they also suggest that some policies have been more successful than others in accommodating the aspirations of minorities. The analysis presented by Mikesell and Murphy creates a formula with which to describe the type of aspiration (from independence in a new state to mere recognition of the group's existence and special attributes within a multinational state – such as the Amish in the US).

At times, the balance of rights for a minority group and the desire to create a sense of national identity in a multinational state may be difficult to achieve. Failure can result in civil conflict on the one hand or cultural genocide on the other. History suggests that nations are not that much more durable than states in terms of existence, as cultural convergence and mixing (as well as outright violent suppression) has eliminated many nations. Language alone is one indicator of the loss of cultures on the globe. Linguists predict that by the middle of the twenty-first century, at the current rate of language loss, there will be only 300 spoken languages on earth, down from approximately 6,000 in the twentieth century.

One driving force behind aspirations for minorities within multinational states to form their own independent political areas may be the perception of uneven development: that some regions are favored over others in terms of investment and economic growth. For example, the recent devolutions of the states of Yugoslavia and Czechoslovakia occurred in part due to separate national identities, but also due to perceptions that regions in each state were not developing at equal rates, or, conversely, that regions developing more rapidly (such as the case of the Czech Republic versus Slovakia) were subsidizing slower growing areas. This issue is even present in Belgium, an apparently stable European state, yet laden with differences among the various national groups, in part reinforced by uneven spatial development and linguistic variations. A basic division between Flemish-speaking and French-speaking citizens of Belgium is reinforced by the rapid development of urban Brussels over more rural Wallonia.

5.6 From tribalism to global village: where is nationalism headed?

This chapter has examined the links between nationality and culture, as well as the way in which we separate ourselves from other human beings and define enemies. On what basis will we continue to identify ourselves within groups? Will citizenship, centered on the state system, continue to be a significant concept to people on the globe?

If we view our spatial identities throughout history as moving from family group to tribal identity, to national consciousness, and perhaps to nation-state, we should then ask whether such a tendency is continuing into the twenty-first century. Perhaps there have always been overlapping identities for people in terms of religion, language affiliation, or cultural identity – "civilizations". The irony of the current geopolitical world may be the two trends that are apparently contradictory yet occurring simultaneously: reinforced and often violent nationalisms side-by-side with an increasingly global scale of identity through cultural invasions (particularly of Westernized culture, helped by monopoly control of technology, the vehicle for such diffusion). For example, Kazakh national identity may be reborn after independence from the Soviet system in the 1990s, but it is an identity framed in part by a now-ubiquitous television system showing Kazakh traditional music alongside advertisements for American candy bars and pirated downlinks from Hong Kong-based MTV programs.

William Pfaff (1993) noted: "Nationalism, of course, is intrinsically absurd. Why should the accident – fortune or misfortune – of birth as an American, Albanian, Scot, or Fiji Islander impose loyalties that dominate an individual life and structure a society so as to place it in conflict with another?" As the organic nature of the world map becomes increasingly apparent, and boundaries form, are broken, and re-form, human beings will need to examine their own perceptions increasingly about their identity on the map: where is our territory? Who are our enemies? What is our nation?

6

Security and Conflict

KEYWORDS

war,
security,
territoriality,
superimposed boundaries,
enemy,
Just War,
pacifism,
place annihilation,
MAD,
first-strike,
proliferation,
terrorism

KEY PROPOSITIONS

- Geography has traditionally been a key variable in warfare.
- Technology changes are reducing the influence of distance in war.
- Prospects of war and the conduct of war create a set of moral choices for society and the individual.
- International borders are increasingly permeable to new types of security threats.

6.1 What role does geography play in war?

More people have been killed in warfare during the twentieth century than in all previous human history combined. Violent conflict is a sign of failure in international and national relations. When a person is killed out of passion or greed, we call it homicide and classify the action as an anomaly, a pathology of behavior, a crime. Yet a large amount of yearly world resources are devoted to training and inspiring people to kill other human beings in a legal and centrally organized fashion. While global spending on the military is actually in the process of declining (see Figure 6.1), some individual states continue to spend massive resources on national defense (see Table 6.1).

Wars are never planned to be long, bloody, and senseless. They are always necessary, a last resort, a test of national will, and envisioned to be short, allowing a clear resolution of international differences. Writer Paul Fussell noted about World War II, "At first everyone hoped, and many believed, that the war would be fast-moving, mechanized, remote-controlled, and perhaps even rather easy" (Fussell 1989: 3). It is always promised that casualties will be minimal, the bombing pinpoint, and the action decisive.

There are many metaphors for warfare. Religious philosophy may tell us that it derives from human sinfulness; psychologists, that it is a type of mass hysteria controlled and manipulated by states and those who stand to benefit from war; strategists analyze it as a form of international relations. War has been viewed as analogous to disease in its spread and effect. It has been likened to natural disasters in its impact on society's struc-

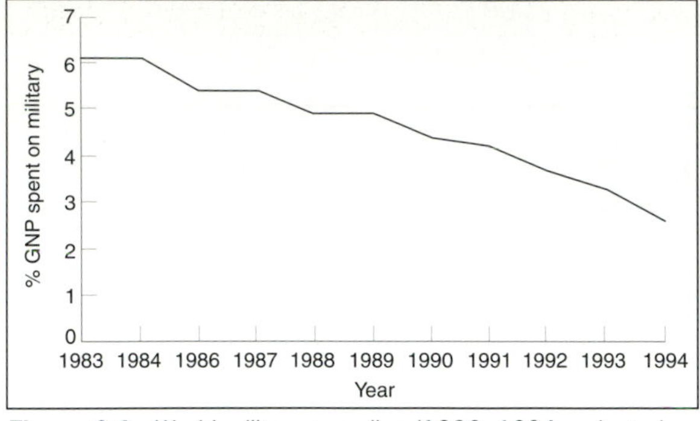

Figure 6.1 World military spending (1983–1994, selected years)

tures. We classify wars, map them, write poetry about them, and build monuments to the dead.

Since ancient times, military analysts have recognized that spatial factors have played a critical role in the resolution of armed conflict. The discipline of geography has much to contribute to understanding questions of warfare. First, spatial, strategic factors are a crucial part of warfare planning and conduct. Military analysts view geography as a key element in the success of campaigns and the ability to understand where a state may be vulnerable. This area is classic geopolitics in terms of analyses such as movement of military

Table 6.1 Countries with military spending above the global average (includes only those reporting)

Country	*% above world average*[a]	*Year*	*Country*	*% above world average*[a]	*Year*
Iraq	70.7	1991	Botswana	2.6	1993
N. Korea	24	1994	Turkey	2.5	1993
Angola	18.5	1986	Azerbaijan	2.3	1993
Russia	18.2	1993	Greece	2.2	1993
Sudan	13.4	1992	Congo	2.1	1992
Oman	13.3	1994	Libya	1.8	1993
Yemen	11.5	1991	Morocco	1.7	1994
Afghanistan	10.6	1990	Singapore	1.5	1993
Kuwait	9.6	1994	Sri Lanka	1.5	1993
Saudi Arabia	8.6	1994	Taiwan	1.4	1993
Israel	6.1	1993	US	1.4	1993
Syria	5	1993	Malaysia	1.3	1994
Albania	4.9	1993	Ethiopia	1.1	1993
Rwanda	4.7	1993	Egypt	1	1993
Laos	4.6	1993	Zimbabwe	1	1993
Jordan	4.5	1994	Qatar	0.9	1993
Mozambique	4.3	1993	Ukraine	0.6	1993
Brunei	4	1990	Tanzania	0.5	1993
Pakistan	3.1	1993	S. Korea	0.3	1993
UAE	3.1	1994	UK	0.3	1993
Vietnam	3.1	1994	Iran	0.2	1993
Bahrain	2.9	1994	Kenya	0.2	1993
Bulgaria	2.7	1993	France	0.1	1993
Djibouti	2.7	1993	Tunisia	0.1	1993
			Myanmar	0.1	1992

Source: Book of the Year, Encyclopedia Britannica.

[a] % equals the percentage of GNP spent on military that is above the world average for that year.

BOX 6.1

Geographers and warfare

The discipline of geography has traditionally been very involved in social issues relating to war and the military. During and after World War II, geographers served the United States military through help with strategic planning, mapping, and area studies. The Office of Economic Research of the Central Intelligence Agency has employed geographers to provide information on competitive states, such as the former USSR. But geographic studies also have provided insights into the impact of the military on regions, including the geographic distribution of strategic facilities and the effects of military spending on local economies. With the advent of nuclear weapons, geographers also turned their attention to examining spatial patterns of potential destruction from a nuclear exchange, responses to government civil defense evacuation plans, voting patterns on nuclear freeze issues, and comparisons of nuclear war and natural hazards research. In 1986 the Association of American Geographers, after much disagreement among members, adopted a resolution asking the United States to take steps to reduce the risk of nuclear war. In his address as President of the AAG called "The Responsibility of Geography", Richard Morrill wrote:

> nuclear war is something apart. Many of our leaders, in both parties, want to believe and want us to believe that nuclear warfare is containable and recoverable. We do not have to retreat from scientific rigor to demonstrate the massiveness and the lingering destructiveness to all forms of life or the futility, the cruel delusion, of plans to evacuate our large cities and survive such warfare . . .Geographers, knowing this, do have a responsibility . . .
>
> (Morrill 1994: 7–8)

vessels through international straits, territorial choke-points, forward capitals positioned for military strategy, and even the distribution of troops and weaponry for battle. Some of the most interesting analyses of campaigns during the American Civil War or the Boer War in South Africa focused on geographic factors. A corollary approach which is inherently geographic is the analysis of a state's industrial resources and capabilities for waging war, vulnerability of trade and supply lines. Governments have often used professionally trained geographers in order to undertake these military and economic analyses. Indeed, the government of Imperial Germany in the late nineteenth century created professorships of geography at every university of the country for that specific purpose.

Second, many conflicts arise from border disputes. Geographers classify and describe various types of borders, and border violations as well may be a matter of geographic inquiry. Boundary disputes are a common theme intersecting geography, international relations, and international law.

Third, the psychology of space may be a realm of geographic inquiry. The importance of territoriality and defending "home" unites geography and psychology. Of particular interest here is research on place annihilation, or the "death" of places during warfare.

Fourth, geographers have recognized that wars do not occur randomly in space around the globe. Distance plays a key role in most conflicts over history, as neighboring states are more apt to war with each other. Yet some areas of the world are especially prone to conflict, as we saw in Chapter 3. Shatterbelts such as the Middle East and Southeast Asia have experienced numerous armed conflicts over the past centuries. Various approaches to analysis have been made in political geography, from behavioral studies to world systems theories about conflict.

In addition to these more traditional ways in which geographic inquiry helps us understand warfare, a new wrinkle on geography and war has been occurring since World War II: the globalization of security threats. This process takes us beyond the realm of traditional borders between states and beyond traditional interstate warfare, to look at the geography of security, not the geography of war. In terms of external or outside threats to citizens of states, borders are becoming more and more permeable despite the best efforts of even wealthy states to protect their citizens. Thus, a new kind of geography, one that looks beyond physical definitions of space to explore perceptual space, and one that moves beyond the state level to global security questions is essential.

Four examples of such "unbounded" threats may be noted already within world geopolitics:

1 Weapons technology changes: as nuclear, chemical, and biological weapons evolve, they overcome space barriers more and more effectively. Even so-called conventional weapons are now reaching a level of technology that will put them on a par in destructive power with nuclear weapons. The technology is driving the need for new theories about geography and interstate conflict as the potential for destruction moves far beyond the scale of national borders.
2 Terrorism: security threats which derive from determined smaller groups, outside the level of regular war, are presented few handicaps by traditional border barriers, yet represent a prime security threat to many states.
3 International crime rings, piracy, and illegal drug trafficking: it is no accident that the United States government speaks of a "war on drugs" and has even given some of the responsibility for fighting that war from civil to military authorities. The United States Chiefs of Staff now include illegal drug trade as a prime security threat to the US. Well-armed and large private "mafias" evolving on the world scene may be the upcoming international security threat of the twenty-first century, and show every indication of an ability to operate quite easily across international borders.
4 Environmental damage: transboundary environmental security risks are becoming recognized as an important element of international relations, conflicts, and legal questions (see Chapter 7).

Plate 6.1 Israeli soldiers on guard at the Tomb of the Patriarchs, Hebron, West Bank (photo: Kathleen Braden)

6.2 Defining concepts of territoriality, conflict, and security

As noted in Chapter 2, power and control of territory are closely related in geopolitics at all geographical scales, and power implies the opportunity to influence others. Alexander Murphy (1990) asked about cases in which historical claims for territory were part of the reason for conflict, and noted that it was not merely the spatial extent of territory that was important in setting off wars, but also the perception of historic claims on territory: do the people of a nation believe they have a justifiable claim to a place? Thus, distance becomes important, since claims are most often made on territories nearer to a state's heartland. Robert Sack has examined territoriality as a facet of human behavior and offered the following definition: "By human territoriality I mean the attempt to affect, influence, or control actions and interactions (of people, things, and relationships) by asserting and attempting to enforce control over a geographic area" (Sack 1983: 55).

Since the geographic pattern of warfare suggests that wars do not occur randomly over space, certain areas seem to be the arena more frequently for conflict over territorial division (see Section 3.9).

A broader view is taken by scholars who explore conflicts at a global level of analysis, often based on Immanuel Wallerstein's research on "world systems" (Wallerstein 1991). This approach suggests that there is regularity in the

Plate 6.2 Rusted tank part left over from the 1967 conflict between Israel and Egypt, Gaza Strip region (photo: Kathleen Braden)

growth and decline of the world economy, set off in part by the spread of capitalism. The globalized economy thus needs a hierarchy to maintain it, and interstate warfare often becomes the prime mechanism. Wallerstein makes the suggestion that a more humane system related to socialism may thus eliminate much of the cause for conflict.

Another world-level approach to understanding war is the theory of the "hegemonic" war, in which the world system is inherently unstable, and subsequent changes (from uneven growth patterns among states) threaten the vital interest of states, and set off conflicts. This idea argues that hegemonic war is one distinct type, and ends up transforming the world state system. The metaphor here is war as disease – that there are symptoms of pathologies in human behaviors and once a war starts, it takes an inevitable course of action.

Classic theorists (e.g. Mackinder and Haushofer; see Chapter 2) who examined the relationship between territory and power were more interested in specific spatial factors. Alfred Thayer Mahan, for example, was a US naval historian in the late nineteenth–early twentieth century who advocated development of sea power for territorial control, noting that states with a good location and favorable coastline were at an advantage for power, although, like Stoessinger,

Plate 6.3 Jammu–Kashmir Liberation Front office in Rawalpindi, Pakistan (photo: Kathleen Braden)

he argued that the nature of the population, national character, and government of a state were also important in the quest to maintain power (Mahan 1890). Air, rather than sea, power was favored by Aleksander P. De Seversky, an American citizen of Russian origin. He wrote in the 1940s that land and sea power were now subservient to air power, and made use of a famous polar projection map to show the proximity of Russia to North America (De Seversky 1950), defining an "area of decision" around the North Pole (see Figure 6.2). We have seen that air power has been an important component of American geopolitics.

All these views have in common the vital link of geography, either as a strategic or as a psychological factor. The American Civil War was, as were most wars of the past, determined in part by the pattern of territorial control. A trade blockade

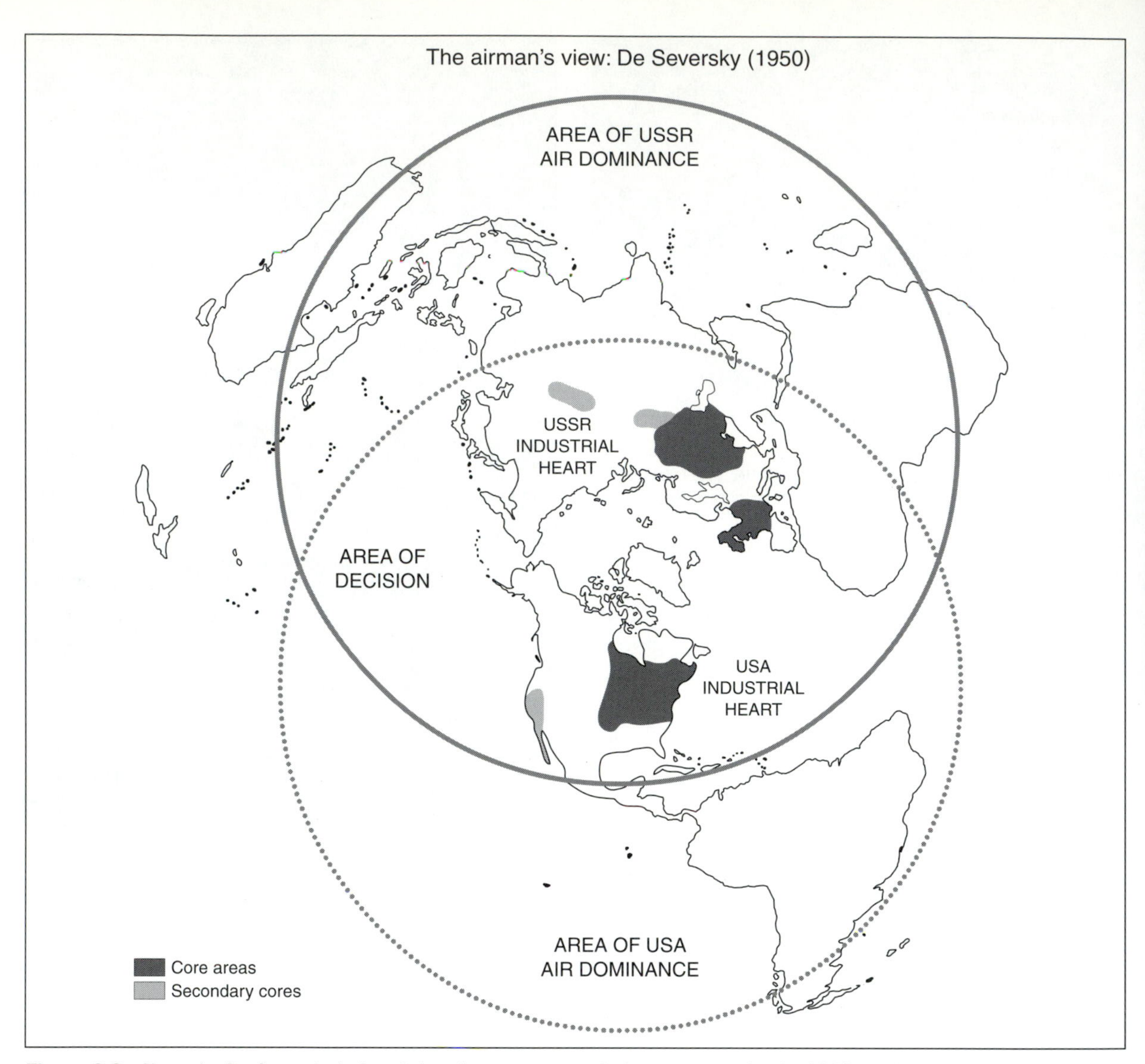

Figure 6.2 Alexander De Seversky believed that air power was replacing sea power by the 1940s and used polar projections to show the US–Soviet face-off across the polar regions. Source: Glassner (1993)

of the South combined with a strategy to cut the South along the Mississippi River, and the psychological warfare of Confederate homeland destruction by General William Sherman helped the Union win the war. The battle of Gettysburg was an important symbolic as well as strategic event because it marked the furthest southern incursion into the northern heartland. Another example is offered by the massive and cruel nature of the geography of Russia, which, coupled with the stubbornness of the Russian army, assisted in various defeats over the centuries from Napoleon to Hitler. On the other hand, the line of glacial hill moraines that extend east–west through the protective marshlands of Belorus proved a convenient "highway" for those who tried to conquer Russia from the west.

What has changed in the latter half of the twentieth century is the role played by geography in conflict. While in any given year, the student of geopolitics may still scan the globe and find numerous examples of traditional conflicts that

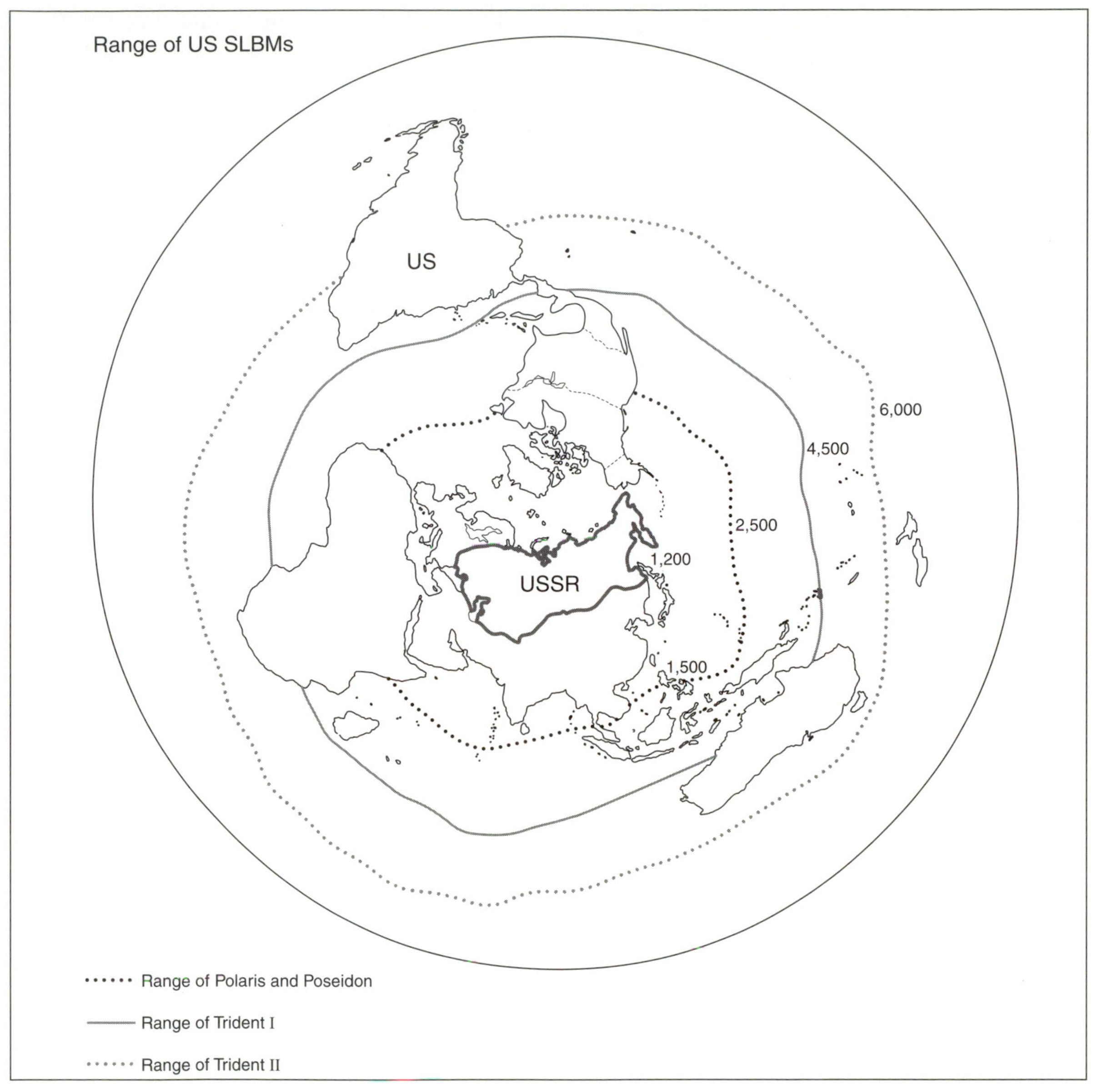

Figure 6.3 The range of US SLBMs (submarine-launched ballistic missiles) has changed with technology. The innermost radius shown is the distance a Polaris or Poseidon submarine could be offshore and still hit targets within the USSR (up to 2,500 miles); the second ring is the range of Trident I submarine missiles (up to 4,500 miles); and the outermost ring is the range of the Trident II submarine missiles (up to 6,000 miles). Technology has thus vastly reduced the role of strategic geography in war-fighting

are framed by geographical constraints, the nature of modern warfare has made traditional geographic theories almost obsolete. At the same time, the psychological and perceptual factors associated with war and deterrence have grown. Thus, domination of the Eurasian heartland whether by air, sea, or land power today may seem meaningless in an era when one commander of a Trident II submarine stationed in the Atlantic Ocean could eliminate every city in the European part of Russia (see Figure 6.3). On the other hand, many argue that the very possibility of such unimaginable destruction has thus far prevented its occurrence.

6.3 War within the state tradition in geopolitics

Three types of traditional wars may be distinguished in geopolitics: interstate wars (one state undertakes warfare on another state); civil wars (the state is a combatant on one side against an insurrection, or two or more groups are battling within a state); and world wars (blocks of alliances are drawn into war, or superpowers engage in warfare). At times, proxy wars in other regions may occur as a surrogate for superpower warfare.

6.3.1 INTERSTATE WARS

Conflicts between states may seem out-of-date in the modern world, and yet when one examines yearly lists of conflicts on the globe, this type of war is still very much with us. Recent examples include the Falkland Island conflict between Great Britain and Argentina (1982); war between Israel and its Arab neighbors (1967, 1973); the horrendous Iran–Iraq conflict (1980–1988); the ongoing Libya–Chad conflict in North Africa; India and Pakistan's off and on again conflict over the Kashmir region; the Vietnam–Kampuchean conflict (1970s). In almost all cases, territorial issues between neighbors were of paramount concern. States located near each other are more likely to go to war than more distant neighbors, and while similar types of states (in terms of social characteristics) tend to cluster, peace and cooperation do not always result. Rather, proximity seems to increase the chance of war.

Murphy (1990) notes that since the state is fundamentally rooted in territory – the idea of place – it is not surprising that wars are the eventual result of unresolved boundary arguments. Despite the fact that the international community has agreed to the principle that territorial acquisi-

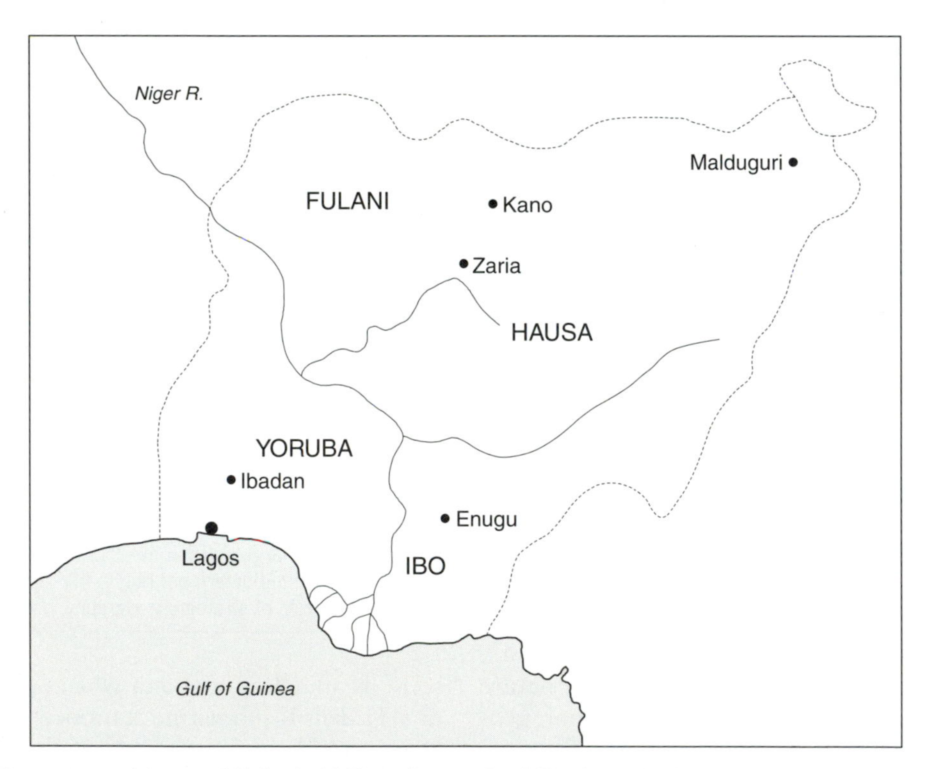

Figure 6.4 The unsuccessful state of Biafra. In 1967 the Ibo people of Nigeria, a mixed animist and Christian group that form one of the largest ethnicities in Nigeria (about 16 million), tried to secede and create a new country called Biafra. Its capital was to be Enugu. The Nigerian government captured Enugu in 1968 and by 1970 the attempt was over, leaving behind up to one million dead, many of whom were children who died of starvation. Source: Chaliand and Rageau (1990)

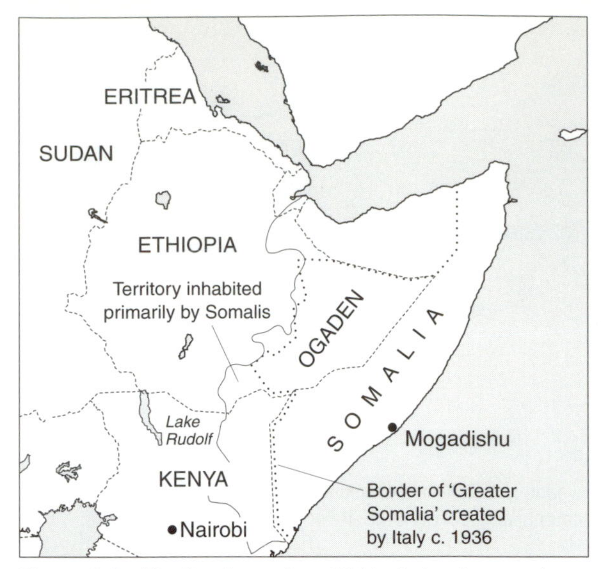

Figure 6.5 The Ogaden region of Ethiopia has been under dispute with Somalia because of the large number of ethnic Somalis who live there. The argument is complicated by the superimposed boundaries drawn not only to separate Italian Somaliland and Ethiopia at the end of the nineteenth century but also between Ethiopia and British Somaliland even earlier. Source: Downing (1980)

tion through invasion is against international law, historic claims on territory are still used as a justification for interstate warfare. Even the infamous 1990 invasion of Kuwait by Iraq was in part grounded in Iraq's ongoing claim to Kuwaiti territory, dating back to the days of British control over both states.

In some parts of the world, boundaries were imposed on local people by outside forces. For example, the division of the non-European world into European colonies resulted in the imposition of boundaries that cut across territories inhabited by distinct nations and ethnic groups. Such superimposed boundaries may create untenable geographic districts in which (1) nations are united artificially and civil war erupts (such as the case of Nigeria in West Africa, which underwent a costly civil war in 1967 when the Ibo people in the east tried to secede and create a new state called "Biafra"; Figure 6.4); or (2) nations are divided by international boundaries and irredentist, interstate wars result (such as the case of Somalia versus Ethiopia over the Ogaden region; Figure 6.5). On the other hand, a lack of any sense of national identity with superimposed boundaries between states may also be a limiting factor on wars.

6.3.2 CIVIL WARS

Wars between groups within a state, often over secessionist movements by a particular national group, seem to be occurring on a frequent basis under the state system which evolved into the twentieth century. Some scholars take this as evidence that states are in essence artificial creations, unrelated to aspirations of nations.

The rights of national groups to self-determination has been offset in the modern world against the rights of territorial integrity and sovereignty for a state. The self-determination for colonial people that led to the proliferation of new states after World War II has been held to somehow be a higher good than self-determination for groups within existing sovereign states. As discussed in Chapter 5, aspirations of minorities for their own sovereign regions may lead to internal wars, or civil wars, between national groups.

One problem that has arisen relates, as noted above, to the problem of defining war. Civil disputes may exist on an ongoing basis for years, but be at such a low level that violence only breaks out occasionally, or an overlap occurs with the popular definition of "terrorism", such as the case of the Basque secessionist movement in Spain or Corsicans against France.

A second dilemma is created by the fact that civil conflicts have been influenced by ideological battles between the superpowers, and often presented through the world press as political struggles, when they were related to inter-group, ethnic rivalries. Thus, the vicious war in Afghanistan between rival tribal and political factions only intensified after the withdrawal of the USSR.

6.3.3 WORLD WARS

This category is used to describe wars that take place between blocs of states, often resulting from complicated alliances which draw in many states to wars that would otherwise have been limited to neighbors, such as World Wars I and II in the twentieth century.

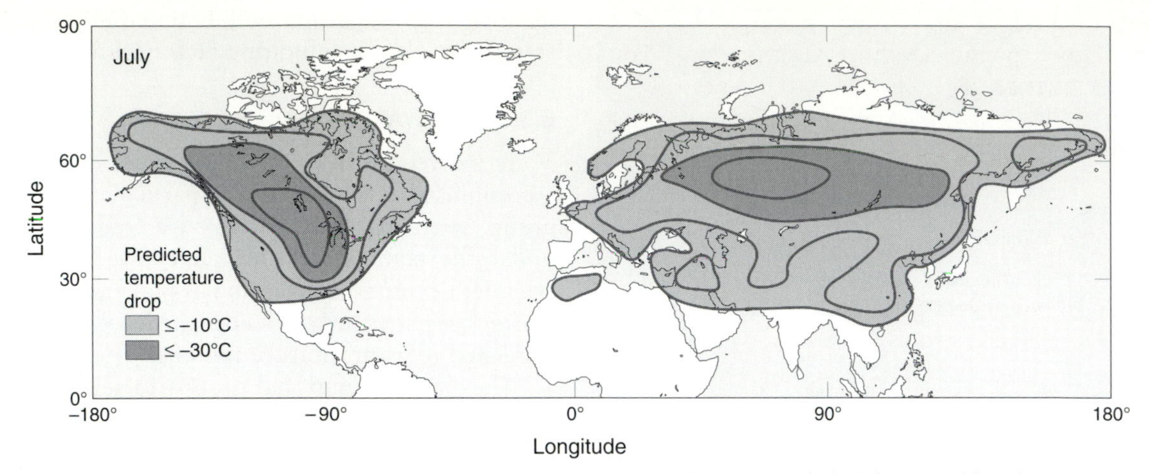

Figure 6.6 It has been predicted that a major exchange of nuclear weapons between the United States and Russia would result in a catastrophic drop in Northern Hemisphere temperatures due to accumulated dust and debris in the atmosphere. Source: Thompson *et al.* (1984)

Structuralist models of geopolitical conflict based on world hegemony aspirations (along with a world systems approach to understanding global crises) suggest that even the civil wars noted above, which took place within the Third World of the twentieth century, may have been a result of economic and political forces unleashed by superpower competition for world power. In some ways, the Cold War was indeed the third world war. Superpower competition to establish spheres of influence reached out to worldwide proportions, with conflict zones particularly in the Middle East, Africa, Latin America, and Southeast Asia, and allied (or even termed "client") states of the USSR such as those of Eastern Europe, Cuba, Nicaragua, South Yemen, Vietnam, Ethiopia, and Angola; with Israel, South Korea, the Philippines, or Somalia, client states of the US. Through military alliances such as NATO (North Atlantic Treaty Organization), the Warsaw Pact, and SEATO (Southeast Asian Treaty Organization), states not directly involved in conflict were drawn into military unions and in some cases into wars themselves.

In actuality, a nuclear war resulting from a major exchange of warheads between the superpowers would have been the ultimate "world war", even drawing in states not aligned to one side or another due to global climate patterns of damage. East and Southern Asian states may have suffered the most severe losses if climate change models were accurate in their prediction of temperature drops due to dust concentrations in the atmosphere. Failure of the Asian rice crop would have resulted in massive famines and perhaps a higher total loss of lives than even direct damage to either the US or the USSR. Thus, China and India, with large rural populations, spatially concentrated and dependent on rice, would be particularly affected by the possibility of a world war (see Figure 6.6).

Perhaps a new level of spatial analysis of the effects of warfare needs to be realized, along the lines of the world systems approach called for by Peter Taylor, John O'Loughlin, and other political geographers. The political and cultural interconnections of the modern world suggest that in an all-out conflict between major powers, the negative impacts and disruption to an intertwined world economy would have repercussions for most states. Even without a nuclear conflict, a new major superpower war would allow few bystanders because of environmental or economic disruption. It may be that the real World War I has yet to happen.

War beyond the level of state blocs may involve clashes of "civilizations" in the twenty-first century. As noted in Chapter 5, Samuel Huntington (1993) suggested that seven major civilization divisions exist in the modern world (Western, Confucian, Japanese, Islamic, Hindu,

Slavic-Orthodox, and Latin American – with African as a possible eighth). Culture-based conflicts in the future could have both global and local implications as competing factions bring their differences across international borders.

6.4 About war and choices

While geopolitics focuses on social institutions, such as delineation of territory, states, and nations, it also examines human actions that are affected by psychology and ethics. There is both a conscientious and a practical reason for looking at values in an examination of war. Even beyond considerations of ethics, national power can be strongly influenced by the ability of a state to rally its people during threatened conflict, and there are documented cases of forces vastly outnumbered and poorly armed who manage to prevail in war.

If we recognize that many wars are started by people in power within state governments, then one must ask whether the state as an abstraction creates and carries out the war or whether citizens are responsible for war. Of course, the answer depends in part on the degree of coercion versus free choice that people are given to engage in warfare, but it may be that "influencing" a citizen to participate in war can take very subtle forms, just as Chapter 5 pointed out the ability of society to reinforce enemy images. When people feel directly threatened and under immediate attack, with their very lives at risk, they may be more willing to support war.

International and national laws suggest that individuals are responsible for knowing about "proper" rules for war behavior. The Nuremberg charter, which evolved out of the war crimes trials held in the German town of Nuremberg following

Plate 6.4 Ban the Bomb demonstration in Washington, DC (photo Kathleen Braden)

BOX 6.2

CO status

CO stands for "conscientious objector", a person who opposes service in the military due to personal convictions. The earliest Christians automatically claimed a type of CO status due to their pacifism, and it was only when Christianity became the state religion of Rome, and soldiers desired to become Christian that ideas of Christians participating in war were accepted. The United States allows for CO exemption under the Military Selective Service Act, which provides for a deferment on the condition that the person performs alternative civilian service and demonstrates religious training or profound beliefs against war. The objection must be universal to all wars. In 1971, the US Supreme Court disallowed exemption based on a person's moral objection only to the Vietnam conflict. The law does distinguish between war and mere use of force under other circumstances. One can receive CO status but still be willing to defend one's family from attack, for example. Deferments are more easily obtained by long time members of certain churches that are recognized as pacifist (such as Quakers) and more difficult if the church subscribes to "Just War" dogma, such as the Catholic Church.

BOX 6.3

Are forgiveness and reconciliation after conflict possible?

The process of healing after warfare is slow and uncertain, yet there are many examples in recent history of a group or a government asking for forgiveness. The idea itself is even controversial, with some people arguing that an official apology from a government for past trespasses is either unnecessary or insincere. However, below are some recent examples of apologies or expressions of regret for violence, wars, or injustices from governments, institutions, or individual leaders.

Date	*Who apologized*	*To whom*	*For what*
1997	Governments of Czech Republic and of Germany	To each others' peoples	German attack on Czechoslovakia; Czech post-war expulsion of ethnic Germans
1997	Moslem leaders of Indonesia	Indonesia's Christians	Attacks on Christian churches by Moslem mobs
Sept. 1997	Labor Party in Israel	Jewish Sephardim of Israel	Discrimination
Oct. 1997	French Catholic Church	Jews of France	Conformism of church leaders during the Holocaust
April 1998	Philippine Catholic Bishops	Filipinos	Failure to support during revolution against Spain in nineteenth century
Jan. 1997	Prime Minister Ryutaro Hashimoto of Japan	Women from Korea forced into prostitution during World War II	Women received monetary compensation from a private fund in Japan as well
1998	US President Bill Clinton	Africans, African-Americans	Regret over slavery
1998	Vatican, Pope John Paul II	Jews, world at large	Vatican non-action during the Holocaust
1998	Government of Canada	Native Canadian peoples	Injustices in their treatment

Equally contentious can be attempts at amnesty, peace agreements, and forms of reconciliation. Guatemala adopted a "Law of National Reconciliation" in 1996 to deal with possible human rights violations during the long civil war in that state, but the law has been controversial; South Africa's Truth Commission to set in motion healing after many years of apartheid laws (separation of races) has been unable to secure the testimony of former president F. W. De Klerk, although he officially repented in 1996 for the apartheid system. In Mozambique, which suffered one of the most horrendous civil wars in recent African history, the opposing FRELIMO and RENAMO parties have tried to build a government together with mixed success.

Archbishop Desmond Tutu of South Africa said, "forgiveness is not nebulous, unpractical, and idealistic. It is thoroughly realistic. It is realpolitik in the long run" (interview with Colin Greer in *Parade*, p. 4, January 11, 1998).

World War II, created certain principles that have become established in international law and in the code of conduct for the United States and other countries. The Nuremberg trials set a precedent that persons who committed crimes against peace and against humanity are to be held individually responsible for their actions. The argument of following superior orders may be relevant to determining the punishment, but does not lessen the individual's responsibility for his or her actions. On the other hand, just being in the service and participating in a war is not a crime. Once in the service, however, a person may refuse an order he or she believes is unlawful. The issue of importance is the degree of choice that a person should be allowed, even during the mayhem of war. The United Nations was overseeing war crimes tribunals for Rwanda and former Yugoslavia in the late 1990s.

What guidance does the citizen receive for making such choices? Religious traditions have much to say about conduct of and participation in warfare, but also have often precipitated conflict in the name of spreading a creed.

6.4.1 PEACE MOVEMENTS

Many social movements in human history have appeared in response to warfare. From ancient Greece, China, India, and medieval Europe to the modern civil rights and abolitionist movements in the United States, non-violence as an alternative to conflict has developed into a strong moral philosophy.

The writings of Dutch philosophers and jurists Desiderius Erasmus and Hugo Grotius in the fifteenth and sixteenth centuries argued for peaceful settlement of differences. Their work established much of the basis for international law on conflict resolution and was heralded as one basis for the United Nations Law of the Seas negotiations. In nineteenth century Europe, anti-imperialist sentiments among socialists and within the labor movement were tied in with calls for peace. The insanity of World War I helped spawn peace societies such as the Peace Pledge Union, the War Resisters League, and the Women's International League for Peace and Freedom.

The advent of the Cold War and nuclear arms race gave a renewed impetus to peace movements, and the debate between anti-nuclear activists and those favoring nuclear deterrence became heated in the 1980s when the United States and the USSR devoted increasing money to nuclear arms. The Campaign for Nuclear Disarmament in the United Kingdom, Physicians for Social Responsibility, and Nuclear Freeze movements were some of hundreds of peace activist organizations that helped bring the issue of nuclear destruction into public consciousness.

6.4.2 WAR AND RELIGIOUS BELIEFS

In the Christian faith, writings and teachings about Christian participation in war present a long, intellectually rich, and diverse tradition. For the earliest churches of Rome, few who were soldiers could also be Christians due to the absolute pacifism argued by church leaders. However, by the Middle Ages, with the church so deeply entwined with all aspects of societal life, doctrine developed which argued that Christians of good conscience could participate in "just wars". St Augustine in his fourth century *The City of God* and St Thomas Aquinas (1225–1274) in *Summa Theologica* created a tradition of examining moral responsibility during war. The idea for having a tranquillity of order and a properly constituted authority led Augustine to argue that a Christian could in good conscience obey if war was unavoidable to restore peace. As later interpreted by the Christian churches, such as Catholic and most "mainline" Protestant, the Just War principles allow church leaders to support Christian participation in war if certain conditions are met: the war must be a last resort; it must be conducted by a proper authority; the cause must be just (aggressive wars are indefensible); there must be a formal declaration of war; intentions must be right; and the ultimate aim must be reconciliation. The Just War tradition was accepted by leaders of the Protestant Reformation as well, such as Luther, Zwingli, and Calvin.

The Crusader tradition (sometimes called, more euphemistically, "the preventative war" tradition because it justifies a first aggression act in war as self-defense) evolved from the Crusades of the Middle Ages. The church during this period, in the guise of bringing the Holy Land back into the hands of Christendom, condoned and even participated in invasion, massacre of civilians, and mass looting. Wars associated with religious differences during the Reformation were among some of the bloodiest seen in European history.

At the opposite end of the spectrum, "peace churches" have held to a pacifist dogma regarding participation in warfare. These include churches of Mennonite, Anabaptist, and Quaker traditions. William Penn wrote in 1684 that governments depend on men, not vice versa, and therefore he went unarmed to sign peace treaties with native Americans. Friends leader George Fox refused to take up arms, noting, "I live in the virtue of that life and power that takes away the occasion of all wars."

The range of Christian moral teachings on warfare was made even more complex with the advent of the nuclear arms race between the United States and the USSR, and the debate about nuclear disarmament that began in the 1980s provided a fertile set of moral and intellectual examinations about individual choice and ethics. In 1983, the United States Catholic Bishops published a pastoral letter in which, although they did not support unilateral nuclear disarmament, they did state that nuclear weapons were morally wrong and deterrence was only an ethical strategy as long as it was a temporary state en route to disarmament. This letter was answered by another group of Catholics led by Michael Novak, who questioned whether this argument was truly within the tradition of the church teaching on maintenance of peace and need for "moderate realism".

Judaism stresses the search for peace as part of God's original purpose for people, but that sometimes destruction is required if it is for a good end. The God who drowned the pursuing charioteers of Pharaoh did so of necessity, but in sorrow. The Talmud distinguishes between optional and obligatory or defensive wars. However, Jews were exempt from fighting on their sabbath when serving in the Roman army. Within Jewish traditions, there is room for a variety of interpretations and moral guidance, and thus much latitude of choice for the individual believer. Perhaps the best word to express the tradition of Judaism is "shalom", meaning peace not just as absence of war, but also as a general condition of well-being, almost analogous to Augustine's tranquillity of order.

Islam is sometimes viewed by non-adherents in Western cultures as a war-like religion, but in actual fact, the teachings of Islam do not argue for war any more than Christian or Jewish traditions, and, as in the other monotheistic faiths of the world, there is a range of interpretations, from the pacifist Maziyariyya sect to the Sufis who argue that necessary war for God's sake must be regarded as spiritual only. The Jihad is the idea of "holy war" in Islam, akin to the Crusader tradition of the Christian faith, and Muhammed the Prophet (who is regarded as the person who brought the word of God through the Islamic faith to humankind) was certainly a military as well as a spiritual leader. However, "Jihad" means "striving", not actual war, and can refer to evangelical activities of the religion which do not require use of force. Laws of warfare under Islam are quite specific, and the attempt to develop a moral code for proper conduct of war resembles later attempts to do so under international laws of the twentieth century and moral codes of the Christian church (such as sparing non-combatants).

The Hindu faith, prevalent in India and other parts of Asia, recognizes a warrior caste (kshatriyas) within its system, but also argues that killing should be avoided if possible. As with other doctrines noted above, rules of war under Hinduism call for protection of civilian populations. Mahatma Gandhi's doctrine of civil disobedience to end British control of India drew great attention to the doctrine of non-violence.

Buddhism, the faith that was originated by Gautama Siddharta in the sixth century out of Hinduism, teaches that one of the "right actions" people must strive for is non-violence. In fact, killing for gain in war, murder, or even killing for food, must all be regarded as immoral. But non-violence must be a personal ethic, requiring meditation and reflection to attain. Despite these teachings, however, examples do exist of states with largely Buddhist populations (such as Sri Lanka and Tibet) engaging in state-run warfare, and some interpret Buddhist teachings to allow for killing if the major doctrines of the faith are endangered.

One religious group that has come on the world scene only recently and has suffered great losses due to adherence to non-violence is the Bahai faith. This religion originated in modern-day Iran in the nineteenth century and offers a vision of a single, unified faith and moral code for the world. Beginning with the 1850 execution of the faith's founder, Bahais have experienced persecution for many years, but refuse a violent response. The present Islamic fundamentalist government of Iran is accused of severely persecuting the Bahais who remain there, but their view on the idea of war and state conduct of armed conflict may be summed up in the words of Baha'u'llah, a leading figure in the faith: "Let not a man glory in this that he loves his country; let him rather glory in this, that he loves his kind."

As long as religion is a principal aspect of human society, it will influence the composition and thinking of the population of a state and the ability of a state to maintain power, and, if need be, conduct wars.

6.4.3 PLACE DEATH AS A STRATEGIC CHOICE

Acts that harm the human spirit and psyche can be a powerful tool during war, but what role is played by destruction of place? When a city is destroyed beyond redemption, what is the exact time when no amount of rebuilding can create the place as formerly understood in its character as home? Because of the particular value of territory to human identity, attempts to kill places, not merely for military purposes, but also as psychological weapons to demoralize the enemy population, have long been a feature of warfare. If there is an image of an enemy person promulgated to inspire people to fight (see Chapter 5), there may also be an image created of enemy place: our landscape is beautiful; theirs, desolate; our buildings are of great cultural value; theirs, a lesser culture.

With the advent of air power in World War I, the idea that aerial bombing of places could be done with "pinpoint" accuracy has become a promise of modern war. While all participating states pledged at the outset of World War II not to bomb cities, by the end of the war, civilian deaths due to aerial bombing of urban places reached into the millions. Many of Europe's old and beautiful cities, including Warsaw, Berlin, and Dresden, suffered extensive damage from bombings during the war.

Many euphemisms are given to describe massive destruction of civilian areas, e.g. collateral damage, reprisals. Aerial bombing during World

BOX 6.4

Describing place deaths

The scene in the animal house is wrenching. A putrid odor pervades the concrete building, and cage after cage is littered with the carcasses of lions, tigers, leopards, and pumas. From the skeletal remains of some and the whole carcasses of others, it is clear that some died sooner than others and that their surviving mates fed on the bodies ... one zookeeper was killed by a sniper while trying to keep up the feeding.

(Article on Sarajevo in *The New York Times*, October 16, 1992)

Vukovar (Croatia) is a wasteland of gutted homes and churches, burning storefronts and bloated, bullet-riddled bodies littering the streets.

(*The New York Times*, November 21, 1991)

It was once so beautiful and now it is all gone. First the government came and bombed us, then the militia bombed us, the mujahedeen bombed us, then the government – everyone bombed us.

(Inhabitant of the village of Paghman, Afghanistan, 1992)

"It was like the moon", said Billy Pilgrim.

(Kurt Vonnegut Jr, *Slaughterhouse Five*, Dell Books, 1969)

BOX 6.5

Walls of sorrow

In Washington DC there is a memorial to the 58,156 American people who lost their lives during the Vietnam conflict from 1961 to 1975. In Novosibirsk, Russia, there is also a monument to the 30,000 citizens of the city who died during World War II from 1941 to 1945. Both monuments make visitors cry.

Aleksandr Chernobrovtsev designed the Novosibirsk monument. It is a series of tall walls, flat pillars, with heroic war images, but also with the image of Mother Russia crying for her children. There are many such monuments in the former Soviet Union, but what is different about Chernobrovtsev's is what is on the *other* side of the walls – the names of every single individual, one out of every ten citizens of Novosibirsk, killed during the war. The architect wrote about why he designed the monument this way:

> I understood that the artist, designing a monument for the very place which had blood shed and lost people, is somehow obligated to immortalize the human tragedy . . . for this reason, I decided that my monument would be to the soldiers themselves, and 30,000 of their names would be placed in one huge list . . . each person is a whole world.
>
> (personal correspondence, May 28, 1990)

While a student at Yale, Maya Ying Lin won an architectural competition for her design of a memorial to the Vietnam Veterans. Two black walls join in the ground near the Lincoln Memorial; the names are engraved chronologically by the date of death, and as you walk the monument, you are drawn deeper into the ground until before you there is just an enormous wall of names and an overwhelming sense of loss.

There are many reasons why societies create monuments to wars – to celebrate particular battles, to show heroes, to immortalize military leaders. They are all built as artifacts on the landscape to help us *remember*. In separate parts of the world, commemorating the losses of two very different conflicts, one American and one Russian architect decided that above all, it is the names that should be remembered.

(photos: Kathleen Braden)

War II was carried out by the Germans against Great Britain, while British and American bombing against Germany and Japan was claimed necessary to save lives later by bringing the war to a quicker end. Even allied commanders admitted, however, that the bombings against cities were ultimately aimed at the terror effect, to break the will of the German and Japanese people.

Geographer Ken Hewitt (1983) studied the aerial bombings of that war and concluded that the spatial distribution of the bombs within urban areas was intended to hit congested, older neighborhoods, often near the center of cities, with better prospects for creating firestorms (conflagrations wherein the very air catches fire and people who are not incinerated are smothered). Major firestorms were set off in Hamburg, Dresden, Kassel, Tokyo, Yokohama, Nagoya, Osaka, and Kobe. Even without consideration of the atomic bombs used against Hiroshima and Nagasaki, civilian deaths in incendiary raids against Japanese cities totaled 780,000.

Threats of place death continued during the Cold War with the deliberate targeting of cities by both superpowers (see Figure 6.7). The idea of civil defense and ABM (anti-ballistic missile) structures was an inherent contradiction in national defense policy. Hewitt notes:

> Urban places and their geography, in particular, are deeply embroiled in preparations for and consequences of war making. There is even a certain direct reciprocity between war and cities. The latter are the most thoroughgoing constructs of collective life, containing the definitive human places. War is the most thoroughgoing or consciously prosecuted occasion of collective violence that destroys places. (Hewitt 1983: 258)

In some cases, places have been killed slowly rather than obliterated all at once. Recent examples include the cities of Beirut and Sarajevo. Regular sniper attacks, bombardments, and supply cuts may create terror among the small remnant population of the city. The psychological effect is to eliminate even that part of the enemy which individual people might otherwise view as surviving: his very culture and civilization, embodied in the built, human environment. Thus, the images associated with places under siege are often remarkably similar in press reports. Beirut after five months of shelling was described as a "moonscape of devastated apartment blocks" (Associated Press, August 16, 1989). Even when cities are rebuilt later, one must ask whether they are the same place, the same idea of home.

6.5 New security threats

Will society have to work even harder to maintain an "us versus them" notion to inspire support for war when the "they" are far away and less identifiable, and the role of homeland, border defense, and territory becomes less clear-cut?

In 1992, the year after the break-up of the USSR, the United States Joint Chiefs of Staff issued their National Military Strategy paper. The Joint Chiefs' document reiterated major "Strategic Principles" for the country: readiness, collective security, arms control, maritime and aerospace superiority, strategic agility, power projection, technological superiority, and decisive force. What was missing was an enemy of equal power to the US. With the apparent demise of the main adversary for world influence, the question was created, just what or who would be future threats to the security of the country?

The doctrine that worldwide "low intensity" conflicts would still challenge the United States led the Joint Chiefs to conclude that the end of the Cold War was no time for the US to let down its guard; rather, the country must be more flexible in meeting new security challenges that required a more diverse strategy, such as the proliferation of nuclear, chemical, and biological weapons, the struggle to improve human conditions, drug trafficking, the march of democracy, and "intensification of intractable conflicts between historic enemies". Thus, the military included even problems of underdevelopment within the security purview of the US, and indeed by the summer of 1994 was serving refugees in Rwanda after a less successful intervention for purposes of peacekeeping in Somalia. Some military officials were beginning to complain that the US armed forces were being viewed as a rapid-relief agency.

According to the document, the biggest threat of all to US security was "the threat of the

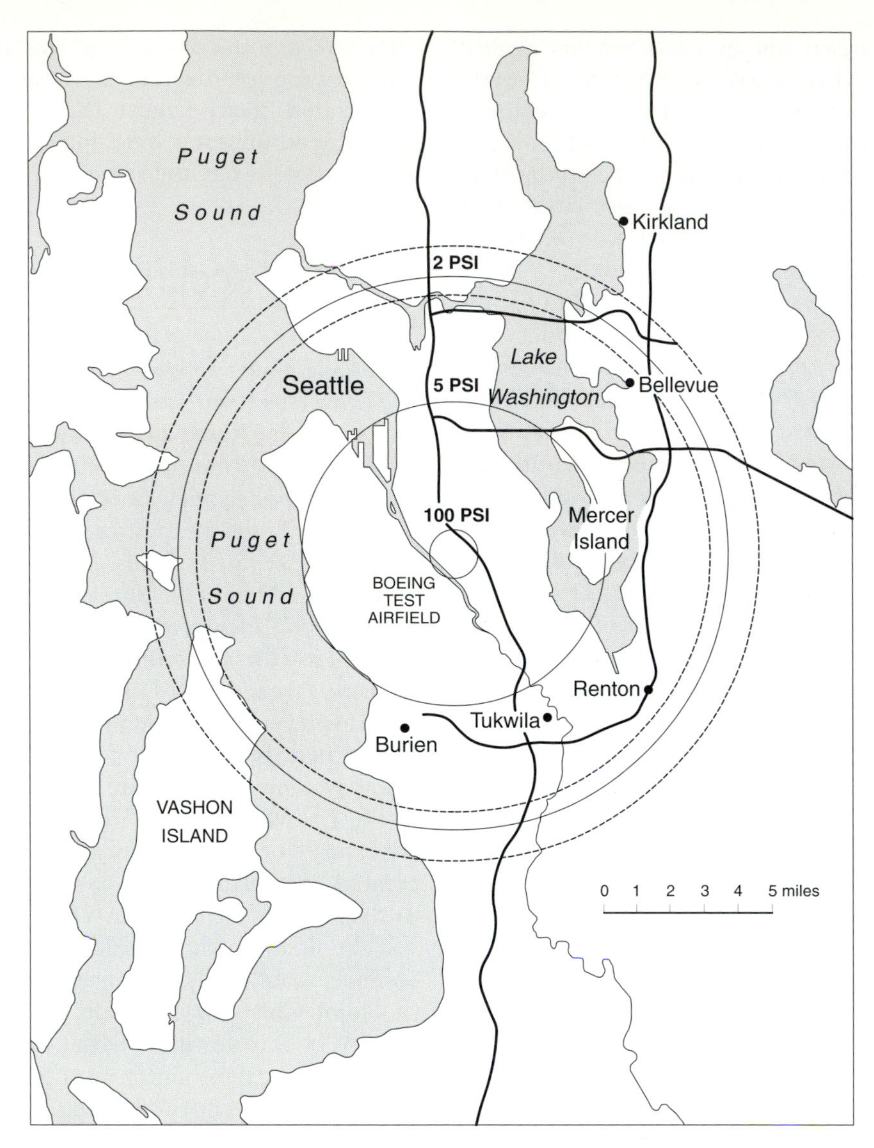

Figure 6.7 The predicted impact of a 1 megaton nuclear bomb on the city of Seattle.
Damage from the blast: damage shown by solid circles. Inner circle (100 psi): all buildings destroyed; no survivors. Intermediate circle (5 psi): all frame houses destroyed; limited chances for survival. Outer circle (2 psi): severe damage to all houses; flying debris.
Damage from heat: the firestorm covers everything within the outer dashed circle, generating winds up to 100 miles per hour. Inner dashed circle: all people exposed to the initial flash suffer third degree burns; survival chances minimal. Outer dashed circle: all people exposed suffer second degree burns requiring medical treatment; all wood, curtains, dry leaves, etc., catch fire

unknown, the uncertain. The threat is instability and being unprepared to handle a crisis or war that no one predicted or expected" (National Military Strategy 1992: 4).

Thus, the business of maintaining state security in the 1990s has apparently experienced a basic shift in nature into something less predictable than the superpower face-off or Third World surrogate-conflicts of the Cold War. General world instability, international crime, environmental problems, and new types of weapons (especially in the hands of non-state agents or "renegade" states) were all noted as threats to the people of the United States. Of interest to the geographer is the fact that most of these threats operate out of the realm of traditional borders. They therefore create a new type of geopolitical security space, which is based more on psychological, perceptual, or jurisdictional space than on defensible physical boundaries. If wars in the next century become harder to classify using traditional definitions of conflict, security threats will increasingly become global and move beyond the control of state borders.

Already there is evidence in many parts of the world that the elite are reverting to systems akin to the Middle Ages, when wealthy persons and small communities established personal protection. Private armies are springing up in the former USSR to protect the wealthy from reprisals by criminals. Multinational firms around the world spend money acquiring technology to help protect their workers against terrorism.

What specific threats to security have emerged recently that deny the principles of geography and borders that have dominated the discipline of geopolitics for so many years? Three are discussed below (weapons technologies, terrorism, and international crime), while the fourth, environmental disruption, is treated in Chapter 7.

6.5.1 NUCLEAR WAR

Atomic weapons have not been used in war since 1945, and yet they dominated the defense of the superpowers during the Cold War and still exist in plentiful numbers in the world. The doctrine of deterrence suggests that a direct war has not occurred because of the threat of mutual suicide. Preparing for war has maintained peace. Yet with the end of the Cold War, new threats related to nuclear weapons have emerged: do North Korea and Iraq have the capability to create nuclear weapons? Will the non-proliferation treaty continue to be honored? Is nuclear "terrorism" a possibility, particularly with questionable control over former USSR scientists and weapons-grade plutonium?

Nuclear weapons are unique geopolitically because they alone offer the possibility of total destruction of the planet's ecosystem. If the nuclear winter theory is correct, an all-out exchange of warheads between the US and the USSR might have caused the destruction of much life on the planet – a type of "terracide". Future regional conflicts in Asia have the potential to cause tremendous damage to the climate of the region, ultimately killing people as much through starvation after loss of crops as from direct destruction.

While the largest arsenals are still held by Russia and the United States, the United Kingdom, France, and China also possess nuclear weapons. The production of nuclear arms is a major industry, and many allied states host nuclear weapons without having sovereign control over their use. There have been almost 2,000 nuclear weapons explosions since 1945, mainly for testing. People who have died directly because of nuclear weapons since 1945 total over 100,000, including people who died at Hiroshima and Nagasaki. Uncounted others may have had their lives shortened due to contamination by nuclear testing.

There are three bases on which nuclear war may differ from conventional, or non-nuclear war:

1 The weapons themselves are much more complex and may even propel strategies for war-fighting than conventional arms.
2 The complexity and cost of the weapons make them prohibitive for development by most smaller states, and thus perpetuates a monopoly situation which has its own irony: superpower blocs alone possess nuclear weapons and yet have chosen not to use them in regional conflicts.
3 Nuclear war must be judged to be separate from conventional war because the outcome would be so distinct. All concepts of defense against nuclear bombs, including the earlier proposed strategic defense initiatives ("space-based defense"), face challenges vastly more significant than those of any past wars.

In addition to these reasons for considering nuclear war as a new notion, traditional geopolitical theories in geography have become less valid thanks to nuclear weapons. The concept of friction of distance is denied not only by intercontinental ballistic missiles (ICBMs), which could have delivered nuclear warheads from the center of the United States to the heartland of the USSR, but also by increasingly sophisticated mobile delivery systems, such as submarines and bombers. William Bunge put the issue thus: "Territoriality has been lost to the species Homo sapiens; we are zero biological distance apart, enemies in a broom closet" (Bunge 1973: 283).

Despite what the world witnessed in Hiroshima and Nagasaki, by the 1950s the population of the West and the Soviet bloc could not be threatened by nuclear weapons on a truly large scale because delivery systems were in their infancy, and geography still played an important protective role, particularly with respect to the US–USSR face-off. The intercontinental ballistic missile, or ICBM, changed this situation and launched the era of massive retaliation. The awesome bombs that exploded in tests so far away in the Pacific could now be delivered right into the heart of the USSR.

By 1961, delivery vehicles caught up with the nuclear explosives technology. The Soviet launch of a space satellite, Sputnik, before the West had managed to conquer outer space, alarmed American citizens. The United States Air Force took the lead in the ICBM program, but the Army and Navy also embarked on nuclear weapons development. In 1960, the first US SLBM, or submarine launched ballistic missile, was deployed on the Polaris submarine.

Although the United States still had a 40:1 lead in nuclear weapons over the USSR, the US population now felt clearly at risk in the event of an all-out nuclear war because Soviet ICBMs could strike the US directly. Civil defense efforts, begun in the 1950s, were increased, and the building of backyard bomb shelters became popular (see Figure 6.8). The deterrence doctrine of civilian populations exposed to nuclear annihilation was termed Mutually Assured Destruction, or MAD. It evolved from the concept of massive retaliation that dated back to the Eisenhower era: any attack

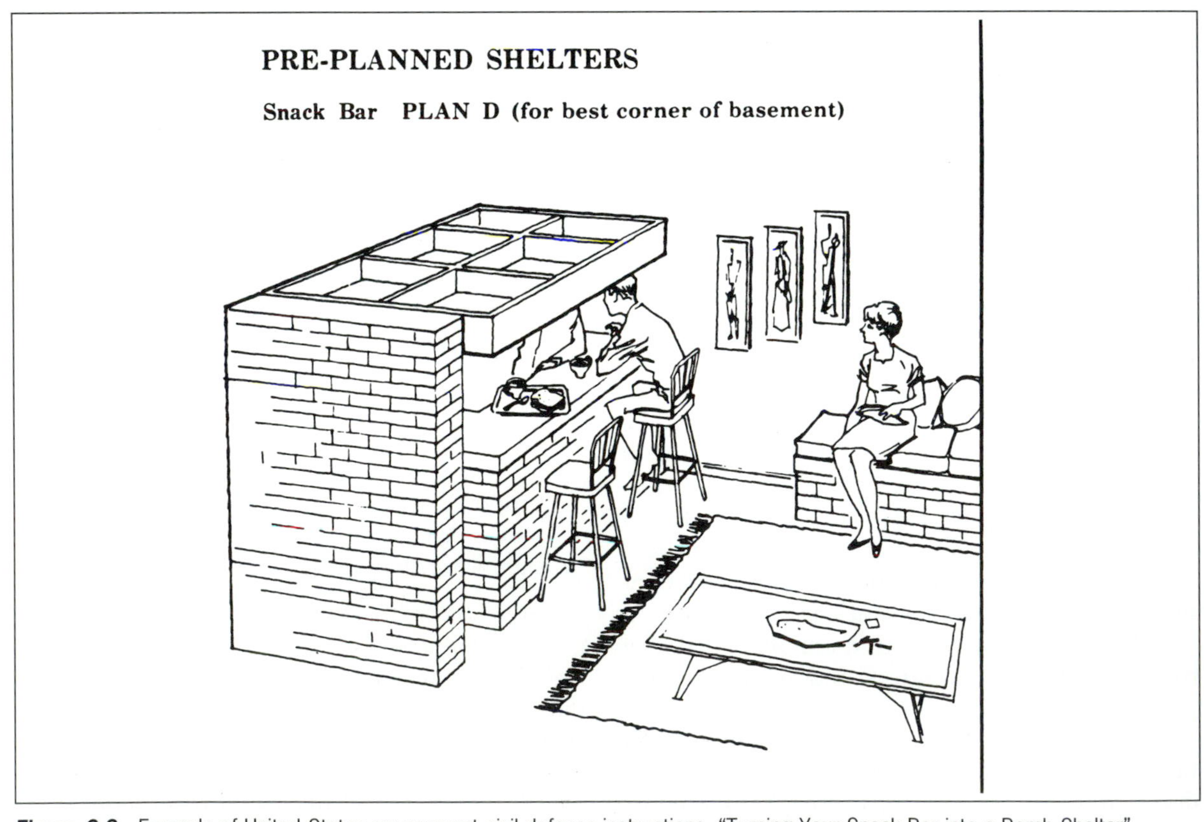

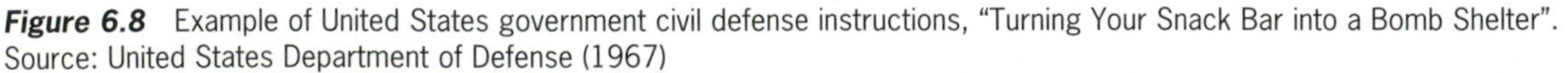
Figure 6.8 Example of United States government civil defense instructions, "Turning Your Snack Bar into a Bomb Shelter". Source: United States Department of Defense (1967)

by the Soviets would be answered by equal destruction from the militaries of the West. In Europe, the doctrine was called "flexible response", meaning that even an invasion by conventional forces from the Warsaw Pact states would be answered by a NATO nuclear response, because the Soviet bloc was believed to have such overwhelming superiority in terms of conventional arms and personnel.

By 1977 a new Soviet missile had appeared that was to cause true alarm in Europe: the SS-20, a mobile missile designed to replace older systems in the western USSR and aimed at Western Europe. NATO (with United States control of the nuclear systems) answered with intermediate-range Pershing II and cruise missiles, and the possibility of nuclear war in Europe was reduced to a seven-minute response time (see Figure 6.9).

Many agreements appeared between the superpowers with respect to arms control in addition to SALT I (Strategic Arms Limitation Treaty, which limited ABM systems): the Seabed Treaty of 1971 prohibited placing nuclear devices on the ocean floor; a multinational treaty against biological weapons was signed in 1972, as well as a US–USSR agreement to oversee naval operations on the High Seas; in 1974 the Threshold Test Ban Treaty prohibited all tests, including those underground, with a yield of over 150 kilotons. The Treaty on Underground Nuclear Explosives for Peaceful Purposes followed in 1976.

An embryonic doctrine was developing, however, during this period which dampened the hopes of halting the arms race. Presidents Carter and Reagan began to shift policy away from MAD and towards fighting a limited nuclear war. While both leaders were interested in overcoming what they believed to be US inferiority to the USSR, Reagan greatly expanded the nation's efforts and developed first-strike capabilities (weapons accurate enough to kill military targets, rather than merely obliterating cities). Thus, nuclear weapons during

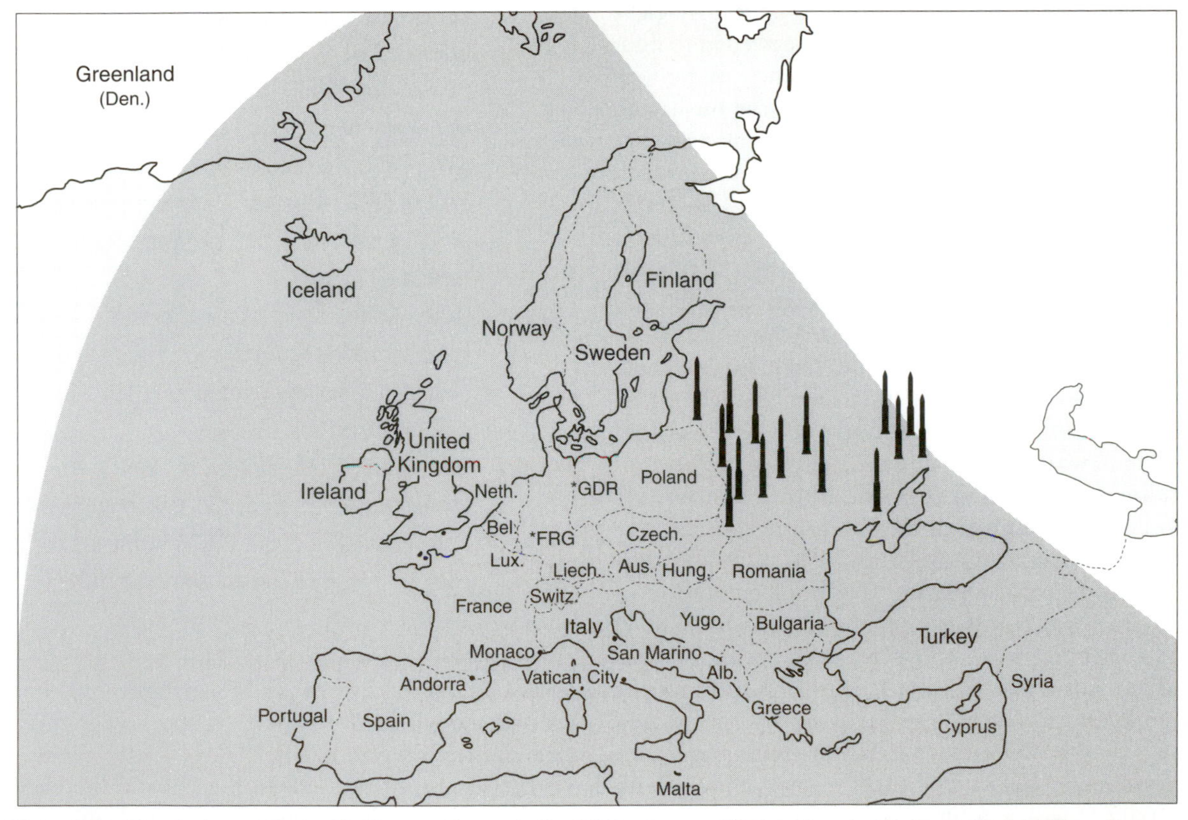

Figure 6.9 The shadow of Soviet SS-20s over Europe in the 1980s. Source: US Department of Defence, *Soviet Military Power*, 1982, p. 26.

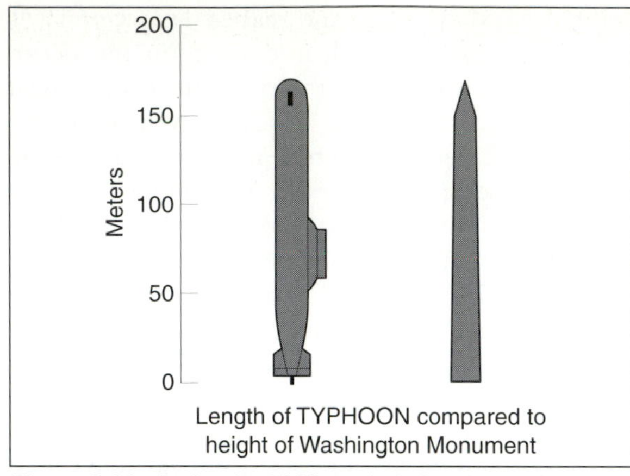

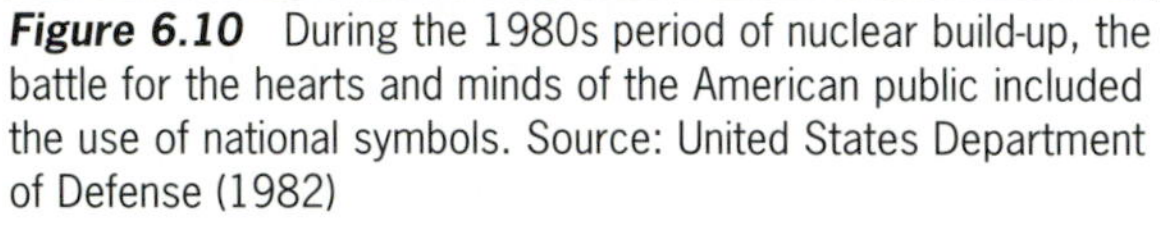

Figure 6.10 During the 1980s period of nuclear build-up, the battle for the hearts and minds of the American public included the use of national symbols. Source: United States Department of Defense (1982)

the 1980s moved away from clumsier, larger, and more numerous missiles to ever smaller, but more mobile and accurate ones. On July 25, 1980 President Carter signed PD-59 (Presidential Directive 59), which discussed a limited nuclear war and the weapons that would be needed to fight one. The Central Intelligence Agency (a famous report issued by Team B, led by George Bush) had argued that the Soviets believed that a nuclear war could be won; therefore, the US needed to match the doctrine with plans of its own. The terms "limited nuclear war", "nuclear war-fighting", "first strike", "counterforce" and "winnable nuclear war" all came into use to describe the departure from deterrence. Under the older MAD doctrine, nuclear war would have no winners. Population centers were to be held hostage to prevent a conflict. The new doctrine suggested that nuclear war could focus on a limited number of military targets and that one of the superpowers could be the winner at the end of the conflict.

The Reagan administration took PD-59 further, and actually planned for implementation of programs to conduct limited nuclear war. NSDD 13 (National Security Decision Directive 13) was issued on November 13, 1981, requiring the Pentagon to create plans for weapons that would be capable of destroying Soviet military targets. Arguments began to surface in public about whether the new administration believed that winning a nuclear war was possible, and images (many involving propaganda cartography) were used to convince the public of the Soviet threat (see Figures 6.9 and 6.10).

Table 6.2 US–USSR or Russian nuclear arms agreements

1972	*SALT I*: Strategic Arms Limitation Treaty (1) Anti-ballistic Missile Treaty (2) Five-year freeze on new ICBMs and SLBMs
1979	*SALT II*: followed informally until 1986. Limits to 2,400 ICBM (inter-continental missiles), SLBM (submarine launched missiles), bomber, and surface-to-air missile launchers
1987	*INF*: Intermediate Range Nuclear Force Treaty aimed to dismantle all Soviet and US medium and short range missiles; inspection by both countries
1991	*START*: Strategic Arms Reduction Treaty aimed to cut the US strategic arsenal by 15% and the Soviet arsenal by 25% within seven years. (US warheads would be cut from 12,081 to 10,395; Soviet from 10,841 to 8,040.) Includes a formula for how many of each type to be allowed
1993	*START II*: United States and Russia set a goal of 50% reduction in strategic nuclear arms by the year 2003; not yet (1999) ratified by Russia
1995	United States again pushing for the International Non-proliferation Treaty to be extended (originally agreed to in 1970 by 172 countries)
1997	New Secretary of Defense William Cohen announces a plan to ask Russia to cut even further, to leave about 2,000 warheads on each side (early Kennedy administration level)

By the end of the 1980s, the US and the USSR together held more than 24,000 strategic warheads. Nuclear weapons had shrunk to battlefield

BOX 6.6

Dismantling missiles

From "All Things Considered", National Public Radio NPR #881028:

> The actual process for dismantling nuclear weapons is classified, but NPR tried to reconstruct the process, using the example of Pershing II missiles placed in Germany. The US Army 56th field brigade maintains these weapons, and each one has a serial number under the nuclear weapons accounting system (of 25,000 weapons). Pershings are kept in several pieces: the missile (delivery system), the radar and guidance system, and the warhead or actual bomb (which has four times the power of the atomic bomb used on Hiroshima, and which is about the size of a trash-can and weighs 200 lbs). The warhead is flown on an air force plane by the "bully beef express" crew to Amarillo, Texas, for dismantling at the Plantex company, which employs 2,750 workers. On an average day, five new warheads are built and five retired, taken apart bolt-by-bolt. There is no chance that a nuclear explosion will occur, but the chemical explosion hazard is risky for workers. Workers take the warhead down into "gravel gertie", a reinforced concrete bunker with two-ton blastproof doors. They take apart electronic components, toxic chemical explosives, and then the radioactive material, which is recycled for other weapons. Ten to twenty pounds of enriched uranium goes to Oak Ridge, Tennessee; five to ten pounds of plutonium to Rocky Flats in Colorado; and a few grams of tritium gas to the Savannah River complex in South Carolina.

size, capable of being moved around by truck, and had grown to projections for basing in outer space. Arms control talks began to be taken for granted, a steady backdrop show to the ongoing military build-up.

However, the disintegration of the USSR (and it is now believed that the Soviet nuclear stockpile has been returned to Russian soil from points in Ukraine and Kazakhstan), pressure from anti-nuclear groups, and massive budget deficits run up by the arms race all created a momentum to reduce nuclear stocks. SALT II, which placed a ceiling on strategic systems, was followed informally until 1986, when the United States exceeded the limit. The 1987 INF agreement (Intermediate Range Nuclear Force Treaty) led to the dismantling of Soviet and US medium-range missiles, which had created the hair trigger situation in Europe. The START (Strategic Arms Reduction Treaty) process superseded SALT and resulted in a 1991 agreement to cut US and Soviet strategic arsenals. START II (agreed to in 1993 but not yet ratified by Russia at the time of writing) calls for a 50 per cent reduction in strategic nuclear arms by the year 2003. The second term of the Clinton administration began in 1997, with Secretary of Defense Cohen calling for even further reductions in warhead levels – down to a number not seen since the Kennedy years (see Table 6.2).

6.5.2 NUCLEAR PROLIFERATION

Will nuclear weapons continue to spread? The main constraints on diffusion of nuclear weapons have been: (1) difficulties in acquiring weapons-grade nuclear materials, uranium-235 and plutonium-239; (2) difficulties in obtaining qualified scientific personnel; and (3) the Nuclear Non-Proliferation Treaty of 1968.

Despite such constraints, several states in addition to the United States and Russia have developed nuclear arsenals, others are suspected of having done so, and still more states are making the attempt. The United Kingdom (who exploded its first device back in 1952), France, and China have independent nuclear arsenals. France decided to construct its own arsenal under President Charles de Gaulle, and first tested a device in 1960. The French have continued to be one of the most stubborn testers of nuclear weapons in the Pacific, despite protests from a variety of international organizations. China first tested nuclear weapons in 1964, and keeps an arsenal of medium-range ballistic missiles, largely as a threat to the USSR. In the 1960s, the US believed that Chinese nuclear arms would soon be as large a threat as Soviet nuclear arms and the US ABM system planned was mainly a response to this perception. However, China apparently has not

Plate 6.5 Militarization of the Sinai border between Israel and Egypt (photo: Kathleen Braden)

developed the strategic (long-range) capabilities that the US foresaw.

India, South Africa, and Israel are thought to be either capable of immediately deploying nuclear weapons, or already compiling stockpiles in secret. India, an early supporter of non-proliferation, obtained appropriate nuclear materials from the West for civilian use under the Atoms for Peace program instituted by the United States. In 1974, India tested what it termed a "peaceful nuclear device". After the test by India, a group of 15 countries, including the nuclear-weapons states, made an agreement in 1978 to be more circumspect about exporting materials and technology that could be used by a country to construct nuclear explosives. India does not claim to have subsequently constructed a nuclear arsenal, despite 1998 tests.

Israel and Africa have entered into an odd type of partnership, which may have enabled both states to construct secret nuclear arsenals. Certainly, the two countries have the scientific technology and labor to build atomic devices. In 1967, after the June War, Israel and South Africa signed secret accords on nuclear cooperation for military and civilian purposes, according to a 1987 report in the *Bulletin of the Atomic Scientists*. Israel's policy has been to state that it would not be the first state to introduce nuclear weapons into the Middle East. Meanwhile, on September 22, 1979, readings from an event in the South Atlantic created controversy over whether a nuclear test had been performed by South Africa, a contention that has never been proved. Speculation about Israel was put to rest in 1986 on October 5, when the London *Sunday Times* published an account from Mordechai Vanunu, an Israel nuclear technician. He indicated that Israel had been manufacturing nuclear weapons for the past 20 years. Vanunu was subsequently extradited to Israel to be put on trial for violating state security.

Iraq, Libya, and Pakistan are suspected of nuclear weapons development. Because uranium used for power production may be enriched and turned into weapons-grade uranium, an attempt by Iraq to build a nuclear reactor was aborted when Israel bombed it on June 7, 1981. Whether Iraq has since embarked on a nuclear weapons program is a matter of debate. The leader of Libya, Gaddafi, was said to be interested in acquiring nuclear weapons, but did not have the materials or personnel to accomplish the deed. However, another Islamic state, Pakistan, is capable of constructing nuclear bombs.

Pakistan has long been a thorn in the side of the United States with respect to non-proliferation. In 1977 the US Congress warned that American aid to Pakistan would be ended if Pakistan continued to develop its nuclear program using French technology, aid was later reinstated when the French halted construction. In 1985, Pakistani agents were caught attempting to smuggle high-technology devices used in nuclear weapons out of the US, and in July 1987 another Pakistani was apprehended while smuggling out special steel for manufacturing equipment used in factories to make weapons-grade uranium. Given the determination of Pakistan to develop nuclear weapons, and the inability or unwillingness of other states to control the diffusion of technology and materials, Pakistan may already have built its own bomb.

Although the cases above illustrate that many states would like to break the monopoly on nuclear weapons held by the superpowers, other states have led the way in refusing to develop nuclear arsenals. Countries which have the technology, access to materials, and scientific community to construct atomic bombs, but who have chosen not to do so include Canada, Sweden, Australia, New Zealand, and Japan, among others. States in the Pacific who belong to an association called the South Pacific Forum signed an agreement to make their region a nuclear-free zone.

BOX 6.7

The boys in the back room

The production of weaponry is big business and very profitable to a number of countries, such as the United States, Russia, France, Great Britain, Germany, and Israel, who lead the world in weapons exports. Technological breakthroughs in weapons development have occurred throughout human history, and warfare has become increasingly industrialized. In fact, William McNeill in his book, *The Pursuit of Power: Technology, Armed Force, and Society Since* AD *1000* argues that only the cost of production was a check on the speed of technical change. "As long as rivalry between mutually suspicious states continues, deliberate organized invention seems certain to persist, cost what it may . . . once again it comes up against the question of consensus and obedience. Material limits are comparatively trivial" (McNeill 1982: 383).

> Those Boys in the Back Room sure knew how to putter!
> They made me a thing called the Utterly Sputter
> and I jumped aboard with my heart all aflutter
> and steered toward the land
> of the Upside-Down Butter.
>
> This machine was *so* modern, *so* frightfully new,
> no one knew quite exactly just *what* it would do!
>
> (Seuss 1984)

The main attempt to halt the spread of nuclear weapons came about in 1968 when the United Nations General Assembly voted to accept a Nuclear Non-proliferation Treaty. Various movements against nuclear weapons had dated back to the 1950s, and different member states of the UN, such as Ireland, Sweden, and even India had offered proposals for non-proliferation agreements. Although 115 states did accept the treaty of 1968, some abstained (Argentina, Brazil, France, India, China, Pakistan, Spain, and Egypt) because the constraints applied only to those states that did not yet have nuclear weapons, not to the superpowers. A provision of the treaty which calls for a halt to the arms race has obviously not been followed by the countries which already have arsenals. In 1995, the treaty was renewed; but India, Israel, and Pakistan abstained from signing the new treaty. In May 1998, both India and Pakistan tested nuclear bombs.

The history of the nuclear arms race shows that nuclear weapons have a momentum of their own. The prevalence of propagandized "gaps" and "windows of vulnerability" on behalf of both superpower blocks suggests that weapons are constructed merely as a response to measures undertaken by the competitive camp. However, the fact is that both sides wanted to build the systems in question and could easily find justification by the real or assumed actions of the other superpowers. The race to build nuclear arms transcends administrations in all the countries which possess them.

By the 1990s, many citizens of the superpower states were breathing easier that the Cold War never became the Armageddon feared for 40 years. Yet the nature of war between powerful and nuclearized states had still been changed forever. Technology had overcome at last the constraints of geography, and new weapons, equally insidious, were on the horizon. What was the lesson learned about the nature of war during the Cold War period: that deterrence was a possible protection against international conflict if the stakes were proven to be sufficiently unacceptable? Or perhaps that people had bought time in the quest to understand, and maybe even control, the meaning of war in the next century?

6.5.3 NON-NUCLEAR WEAPONS TECHNOLOGIES THAT DEFY GEOGRAPHY

In his examination of changes in weapons technology over human history, William McNeill begins with a look at bronze weapons, which were introduced to Mesopotamia in the fourth millennium BC, followed by the use of chariots and cavalry horses. As with many weapons innovations, the factor that ensures their efficacy in war is not so much the destructive power of the weapon, but the delivery system capable of bringing it to the enemy. Thus the movement of weapons over space, a basically geographic notion, has always had much to do with their success.

Nuclear weapons became of strategic threat to opposing superpowers, not when the hydrogen bomb was developed, but when intercontinental ballistic missiles, submarines, and long range bombers allowed their delivery right into the heartland of the enemy. Deterrence was in part dependent on the ability of nuclear weapons to overcome the friction of distance between enemy states.

Conventional weapons in the US and other states' arsenals are now also quite capable of overcoming vast distances for delivery, although the ability of naval carrier units and airplanes to deliver the weapons is also an important factor. Ideas have emerged, such as the neutron bomb (where people are killed by massive radiation but property is protected from blast effects) and the Strategic Defense Initiative, to move defense to the outer space level. Chemical weapons are one of the older type to defy geographic control because they rely on poison gases or other deadly substances spreading through the air. In fact, at the battle of Loos during World War I, British use of poison gas literally backfired as a wind shift sent the gas back upon their own troops.

Fifty different types of gases were used during World War I, and in 1925, a protocol was adopted in Geneva to ban the use (though not the production) of chemical weapons. However, they have continued to be used in this century, in part, because the technology is simple and their production is not expensive. The United States, USSR, France, and Iraq have all continued to stockpile chemical weapons, but other nations, such as Libya, are suspected producers; accidents have poisoned terrain and killed people and livestock, and their use in warfare during Iraq's suppression of the Kurds, the Iran–Iraq conflict, and the USSR's war in Afghanistan have all been reported. Likewise, US use of chemical defoliants during the war in Vietnam has been argued as a severe health problem for US veterans and citizens of Vietnam. In 1990 the US and the USSR agreed to begin destruction of existing stockpiles of chemical weapons, with 50 per cent of national stores to be destroyed by 1990.

In 1993 the Convention on the Prohibition of the Development, Production, and Stockpiling of Chemical, Bacteriological, and Toxic Weapons was signed by 143 states. By 1997, enough states had signed and ratified the convention (including United States Congress) to put it into effect.

Earlier (1972), many states had signed the Biological and Toxin Weapons Convention against their stockpiling and production. Anthrax, botulism, plague, enterotoxins, and saxitoxin substances have all apparently been subjects of experimental weapons development, and the US admitted to having a developmental program until the 1960s. An accidental release of anthrax toxins apparently happened in Russia's Ural Mountains, contaminating the Sverdlovsk area during the late 1970s, and the event became the subject of an infamous cover-up. The release of germs for war-fighting may prove the ultimate weapon to overcome the constraints of geography; unfortunately, the ability to then control the spatial spread of effects may be an illusion.

6.5.4 TERRORISM

As with the term "war", the word "terrorist" is difficult to define because the violent act one group regards as terrorism may be akin to an act of patriotic warfare by another. It is basically an event of violence or threatened violence that moves outside the realm of the state system, but is often directed at a state. The following elements help to define the concept of terrorism:

- political motivation, rather than economic gain;
- threat of violence, particularly of a "surprise" nature and directed at innocent bystanders or civilians (although military targets have been picked);
- maximum use of publicity to draw the world's attention;

- committed by a group, not part of a sovereign state government, although this factor is a matter of controversy. The United States government claims that certain states may be identified as direct supporters of terrorism.

By its nature, terrorism is a clandestine activity that occurs beyond the constraints of borders and distance, despite efforts to protect citizens through such devices as metal detectors. Once the terrorist act occurs, publicity value is important, therefore, some experts have suggested that press coverage be limited so as not to provide a forum for terrorists. Such a procedure, however, runs into freedom-of-the-press provisions in many states.

A policy to never negotiate with terrorists has also been suggested, but political realities mean that this would be a difficult plan to implement, even in Israel, which had such a policy for many years. One form of terrorism, airplane hijacking, has been made an international crime via the 1970 and 1971 conventions against hijacking.

Europe has been an especially prime location for terrorist attacks due to its geographic position as a world air travel crossroads, the number of nationalist disputes within Europe, and past colonial ties to Third World states. The opening of borders with the European Union plans have made police wary about their ability to provide security within Europe as goods and people circulate more freely.

Can there be a military solution to terrorism? The 1993 bombing of a high-rise building in New York City proved that even the United States may not stay immune to such security threats. In the past, the US has been willing to use military force to answer terrorist attacks, but such actions require reliable intelligence information about the perpetrators and an interpretation of terrorism as a direct act of aggression. With greater potential for acquisition of even nuclear weapons know-how and nuclear materials by terrorist organizations, the ability of states to provide traditional defense against these threats may be diminishing.

6.5.5 ILLEGAL DRUG TRAFFICKING, INTERNATIONAL CRIME

The US government is increasingly making drug trade a national security issue, moving it from the realm of civil crime into military purview, and the possibility that international crime rings may now be dealing in nuclear and other advanced weaponry items is increasing the fear. In addition to drug trafficking, international crime groups have also affected states' security in absolute disregard for traditional border defense in areas such as money laundering, illegal immigrant traffic, and even the growing black market for human body organs thanks in part to technology for transplant operations.

The breakdown of borders in Europe following the European Union agreements and the demise of the tight security control under the USSR have added two elements that have facilitated the international flow of illegal activities. Many of the crime groups are related to major ethnicities (Sicilian, Russian, Japanese, Chinese) and therefore can take advantage of family-like connections as well.

The Pentagon now includes the "Drug War" within its budget lines, in an attempt to provide a traditional military response to a security problem that represents the new "borderless" threat to citizens. Arguments are even heard that the Cold War itself facilitated the development of international narcotics trafficking because drug runners were often working in alliance with covert operations abroad, funded by the superpowers, and drawing much of the financial support needed for national insurrectionist movements (such as in Myanmar and Afghanistan) from the sale of narcotics (see Figure 6.11). Military counters to drug trafficking have involved attempts at interdiction, intelligence gathering activities, training and arming of foreign troops to counter narcotics crops at the point of origin, and even massive aerial spraying campaigns to try to eradicate drug plants. Despite all these efforts, liberalization of world trade and borders has allowed illegal drug trafficking to continue, and threaten states from within thanks to crime and disease problems. The advent of containerized cargo has even facilitated movement of narcotics, making illegal drug shipments harder to detect.

Finally, the world may now be experiencing the melding of two categories of trans-border threats as international crime cartels engage in the sale and illegal transport of advanced weapons technology.

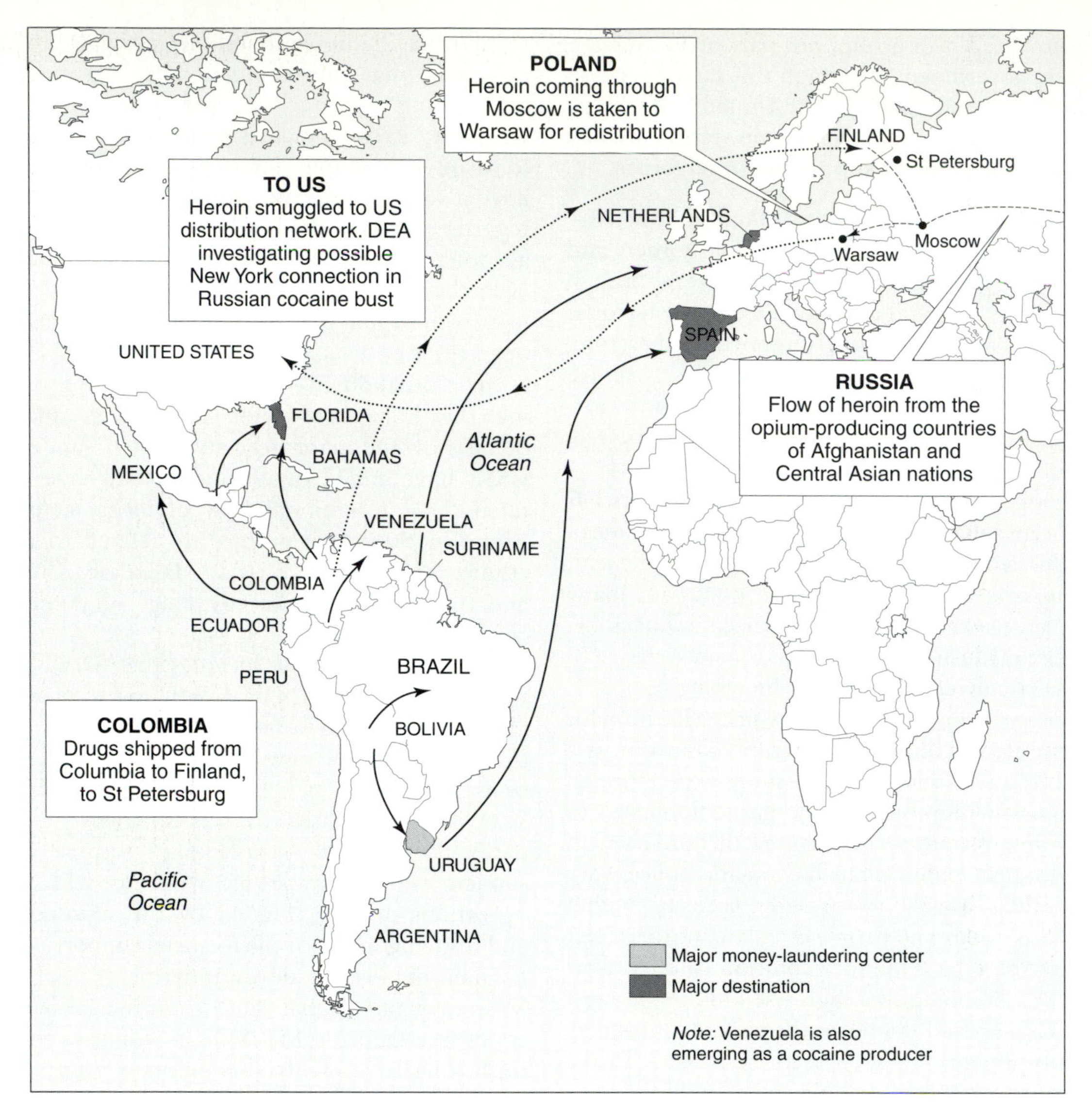

Figure 6.11 Assumed cocaine and heroin routes into the United States. Latin American and European states as well as Russia are believed to serve as conduits for illegal narcotics entering the US, as points for either transhipment or money laundering

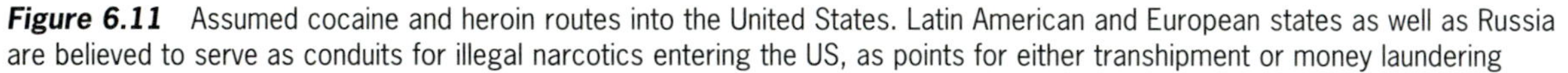

In August of 1994, Germany arrested two Spanish citizens and one Colombian citizen on their way back from Moscow. They were transporting 17.5 ounces of weapons-grade plutonium, and this was the third instance of Germany confiscating illegal plutonium en route through its borders.

The economic crisis in the former USSR has impoverished many scientists and workers with access to advanced technologies and products.

While traditional military protections may be of minimal value in overcoming these new threats to security, several trends can be noted which may help formulate future security policies:

1 The international, trans-boundary nature of serious security threats in the twenty-first century needs to be recognized by states' goverments who may otherwise attempt to develop "bandaid" traditional solutions, believing that borders are still defensible.
2 Security threats now have much more of a collective nature and therefore require collective

BOX 6.8

Mauritius: a multinational state not at war

Mauritius is an anomaly in a world of conflict, a state of several small islands in the Indian Ocean off the coast of Africa with a very diverse population apparently at peace with each other. The official language is English, since the last in a series of colonizers were the British (after the Dutch and then French took the island), but the island became independent in 1968. The population is divided between African–European creoles, Indians, Chinese, French, and other Europeans. Hinduism, Christianity, and Islam are all practiced freely on the island, an area of 2000 km^2 with a population of slightly over 1 million. Thus far, this diverse lot of people have managed to live together peacefully within a democratic state where the main challenges are building an economic base and keeping an eye on a high birth rate. Mauritius is not exactly an island paradise nor perhaps a model for world cooperation, but it does manage to keep peace with only a small police force and minimal military expenditures.

solutions and more international cooperation among governments who serve states and international agencies.

3 Intelligence gathering seems paramount since scarcity of information increases the risks.
4 Negotiation with political groups engaged in violent activities may be required in recognition of minority group aspirations.
5 Development of advanced weapons technology is still largely an activity carried out by wealthy states for profit or perceived extension of national power. Therefore, the responsibility for their potential use against nations must be shared when illegal groups or "rogue" states obtain the technologies which might otherwise be too expensive to develop.

At the start of the twenty-first century, conflict and security matters for states will become more complicated and traditional defenses less effective. Geography is proving a diminishing barrier to a host of security breaches, which may affect citizens even of wealthy countries. Change is then fast affecting the twin notions of sovereignty, which depends on territorial control and power, which allows control of other groups' actions. In short, upcoming security threats may indicate that states no longer have the luxury of defining an "us versus them"; in a world where borders are increasingly ineffective, everywhere is home territory and we are indeed enemies in a broom closet.

7

The Environment

KEYWORDS

biosphere,
ecosystems,
resource,
fleeting resource,
value,
scarcity,
externality,
common property resource,
transboundary impacts,
renewability,
sustainability,
biodiversity,
UNCED,
GEF,
tradable emission credits

KEY PROPOSITIONS

- International security is increasingly dependent on a healthy global environment.
- Externalities, determination of liability, and compensation systems require a developing body of mechanisms to solve international disputes.
- Sovereignty of states over their environment is gradually giving way to global tools.
- Non-governmental interest groups are playing an increasing role in environmental issues.
- North–South inequities exacerbate environmental problems globally.

7.1 The intersection of environmental concerns and geopolitics

In the mid-fourteenth century, France and England began a transition from feudal kingdoms to centralized monarchies. Thus, the modern state system began to coalesce in Europe, in part to increase security to citizens. Yet by the end of the century, one-third of the population of Europe had perished. The bubonic plague had entered the continent, carried by fleas on the rats that inhabited the trade ships from the east. No existing tool of state security could protect the citizens of Europe from an unhealthy environment.

Today, a major portion of international relations deals with protecting the earth's environment. As we understand increasingly that resource use and degradation of the planet are phenomena that rarely stay within international borders, it becomes apparent that issues of the environment may be at the very core of geopolitics.

But what is the environment exactly? The Western, neoclassical view of the environment and of natural resources seems increasingly to dominate international discourse about the environment. It includes the following elements: first, that there is a dualism between people and nature; thus, we say natural resources to distinguish them from, for example, human resources. Second, discourse is generally grounded in neoclassical economic theory, which commodifies elements withdrawn from the earth and talks of their utility to society for consumption, economic growth, and ownership.

More specifically, if we are to speak of the earth environment with all its living organisms and the relationship between all its components, the word biosphere may be more appropriate. Scientists began using the term in the nineteenth century, and the work of V. I. Vernadsky in Russia brought special attention to the idea in this century. Vernadsky argued that the agency of humans in modifying the biosphere actually replicated a type of geologic modification force on the earth.

Below the planetary level, biomes can refer to large systems that are often represented by major vegetation types, but which also include the interactions of all the actors within. The science that allows the study of these interactions is termed ecology and to pluck out a sample of interactions at any scale could be to delineate an ecosystem at work. The transfer of energy, beginning with the sunlight received by plants, may be the highway structure that allows us to describe ecosystem activities.

In Chapter 6, we noted that the threat of nuclear war increased as the United States and Soviet Union developed the technology to deliver nuclear weapons quickly over long distances. The world has thus "shrunk" and more sophisticated weaponry coupled with the globalization of the world economy has meant that the scale of human activity is moving to the biosphere level, much along the lines of Vernadsky's thesis. Boundaries have always been a difficult concept to apply to ecosystems: geese migrating between Russia and the United States may transcend ecosystem boundaries at several levels; their very flight creates a new scale of ecosystem. Add to this the complication of separate hunting regulations in different states, within human-made boundaries not recognized by the geese, and we have a recipe for a problem in geopolitics.

The idea of environment, therefore, in this chapter, refers to international relations and conflicts that arise at the global level (biosphere-impacting) or regional level among two or more states (regional ecosystem-impacting).

While people typically think of wood, metals, plants, animals, and other elements of the natural world as resources, in actuality, the concept of "resource" demands human agency. Natural substances are only resources if extracted and used by human beings. The understanding of resources and questions of international cooperation for resource management demand an appreciation of the concept of value, albeit recognizing that values may vary among places and cultures. We say that we can "use" resources, but we are in essence modifying the elements of the biosphere to increase their utility to us.

"Fleeting" or mobile resources are especially subject to geopolitical conflicts because of their potential to transcend boundaries: fish stocks, pools of underground oil, migratory fauna, and the fleeting character of resources used up to the point of noxiousness, e.g. the transformation of air and water vapor resources into acid rain, which then moves across international boundaries between Eastern Europe and Scandinavia or between the United States and Canada.

What is the correct way, then, to use an earth element? Is it acceptable to use resources to the point of exhaustion? Do individual states have the right to make those choices within their own boundaries? These questions are difficult to answer for overtly tangible resources such as uranium or forests, and are equally challenging for resources such as fresh water or clean air.

7.2 Resource concepts applicable to geopolitics

As the world seems to be adopting the Western model of resource consumption and utilization for economic growth, concepts long rooted in neoclassical market economics are spreading globally. In historic terms, the Western, industrialized states have been the big consumers of the world's resources. In the future, however, developing states will play an increasing role in impacting the earth ecosystem due to population growth and increasing levels of material consumption, repli-

BOX 7.1

Geopolitical conflict and the environment: a fish story

Our fish story demonstrates the importance of bringing together resource management interests, proper economic tools, and international perspectives to solve transboundary resource issues. When such action cannot be accomplished, sustainability of resource utilization is endangered and relationships between states can sour.

The fish in this case are salmon, born in rivers of North America but spending much of their life out on the Pacific, to return for river spawning and forming a multibillion dollar resource for the people of the Pacific Northwest. Their lifecycles transcend geographic boundaries and create for the people of Canada and the United States a common property resource management dilemma. All stakeholders in the resource's continuance must share in ownership responsibilities or the renewability of the resource is threatened. In turn, any one stakeholder's actions can have impacts on all others.

The stakeholders are as follows:

1 the governments in the region (US and Canada at national level; British Columbia, Washington, Oregon; municipal governments of local fishing towns)
2 private fishers who make up commercial fishing fleets
3 Indian tribal groups
4 sports fishers
5 commercial distributors and consumers

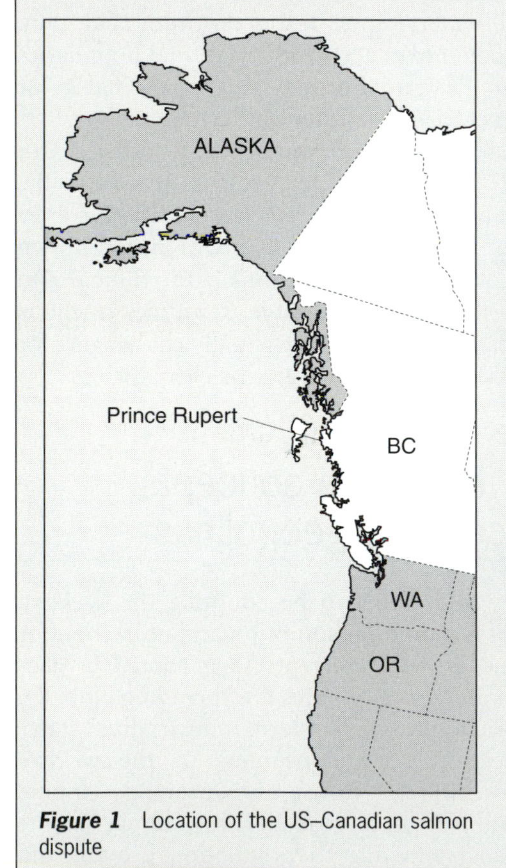

Figure 1 Location of the US–Canadian salmon dispute

The geography of the region complicates management in that salmon go to the panhandle of Alaska and then swim south to BC waters (Figure 1). US commercial fishers wishing to work off the coast of Alaska (in US territory) must ply through territorial waters of Canada.

Due to an obvious decline in salmon, the United States and Canada signed an international treaty in 1985 to specify rights and responsibilities. The treaty guaranteed a catch for each country "equivalent to the production of salmon originating in its waters". But British Columbia and Alaska have experienced difficulties in agreeing about equitable catches. Canada accused Alaskan fishers of taking millions of dollars in excess of their allowable catch yearly and offered to unilaterally cut its share of chinook capture by half to reduce pressure on the common resource, but Alaska refused to reduce its catch below a threshold. The governor of Alaska said that overfishing was occurring on the BC side and that Alaskan fishers had already been reducing their catch.

Meanwhile, a committee established to manage chinook salmon recommended a 50 per cent reduction in harvest all around to save the stocks. When no agreement was reached, Canada began to retaliate and imposed a $1500 toll on US fishing boats traveling through BC waters. To further complicate the geopolitical relationships, Washington and Oregon began to side with British Columbia in the dispute against Alaska. An international mediator was called in, but quit when the sides could apparently not be reconciled. Then, in June of 1996, 24 Indian tribes reached an agreement with the governments of Washington, Oregon, and Alaska over the Southeast Alaska level of chinook harvest, pegging harvest rates to predictions of stock levels rather than a fixed amount. However, this agreement was not extended to an international convention with Canada, and Canada is still unhappy with the level of fish Alaska was allowed under the tribal treaty.

cating Western growth trajectories, but at a larger scale and with increasingly scarce resources. The pattern thus emerging suggests that geopolitical conflicts over resources will increase.

7.2.1 CONCEPTS RELATING TO RESOURCE QUANTITY AND UTILIZATION

Classical Western economists, such as John Stuart Mill and David Ricardo, developed the notion that resources are inherently scarce (i.e. limited in quantity and location). Competition for limited resources would therefore be a check on economic growth of states. Two examples are Japan's rapid colonial expansion during the first half of the twentieth century in search of resources to fuel its developing industries and Great Britain's colonial extension, in part due to depletion of industrial resources in Europe.

Resources may also be divided into those that are renewable with proper management tools and therefore can be utilized on a sustained basis over a specific time horizon (fish, wood, clean air, fresh water) and non-renewable, with a fixed supply that may run down within a human time frame (fuels such as petroleum and coal, minerals). But renewability may be a subjective concept. Timber, for example, is regarded as a renewable resource because trees can be replanted, but plantation management of timber usually results in a monoculture that is far from nature's view of what constitutes a forest. On the other hand, human ingenuity has proven formidable when supplies of resources become quite scarce, forcing the search for economically viable substitutes. Yet the soil, air, and water of the earth are proving to be indeed non-renewable resources if damaged to the point of non-restoration. United Nations studies in 1992 and 1995 respectively indicated that approximately 89 million hectares of land on earth are so damaged by improper utilization that they cannot be reclaimed and that fresh, clean water is becoming more scarce, with 80 countries facing severe water shortages in the near future.

World consumption patterns of many resources show that the traditional heavy consumers of resources have been the wealthier, industrialized

BOX 7.2

Antarctica

Antarctica, with its area of 14.2 million km^2, is 40 per cent larger than the United States, but it has the distinction of being an area with no permanent settlements and no one state sovereign over its territory. Yet it is a land of potentially wealthy resources and has therefore become an actor on the world stage of environmental geopolitics. Several states have claims to portions of Antarctica, including the United Kingdom, New Zealand, France, Australia, Norway, Chile and Argentina, but other states, including the United States, Russia, Japan and Brazil, also have interests, largely connected with scientific research. An agreement made in 1959 held that human activity here should focus on international cooperative research and no military uses of Antarctica are allowed. In 1964, Antarctica was identified by a convention as a Special Conservation Area, designed to protect its flora and fauna, and conventions on seals and marine living resources followed in 1972 and 1980, respectively. But human disturbance of the pristine Antarctic environment due to research stations and increasing tourism have raised concerns over better protection. Perhaps the thorniest international issue, however, is the regulation of mineral exploitation. Coal, oil, natural gas, and many metallic minerals including uranium, gold, silver, and titanium have been located in Antarctica. A convention was refused by Australia and France in 1989. A protocol on environmental protection was accepted in 1991, but issues of enforcement and liability for damage were still open questions.

states. These countries have not only used their own resources, but also have extracted large quantities from developing countries (see Figure 7.1).

Yet, as seen in Figure 7.2, the share of world energy used by Western states has been declining, while developing states (especially the larger ones such as China and India) have been increasing. A comparison of energy consumed by the United States and Thailand for the years 1983 and 1991 demonstrates this trend (see Figure 7.3). Thus, countries moving into the NIC (newly industrializing country) category are the consumers of the future, leading to questionable sustainability of the resources. The West, meanwhile, has been becoming more energy efficient, but then, it can afford to be so.

The World Resources Institute notes that the world is not likely to run out of non-renewable resources within the next few decades, and in some cases, prices for resources are lower now than 20 years ago as more countries enter world resource markets. However, prices may not always reflect the total social value of a resource under existing pricing tools.

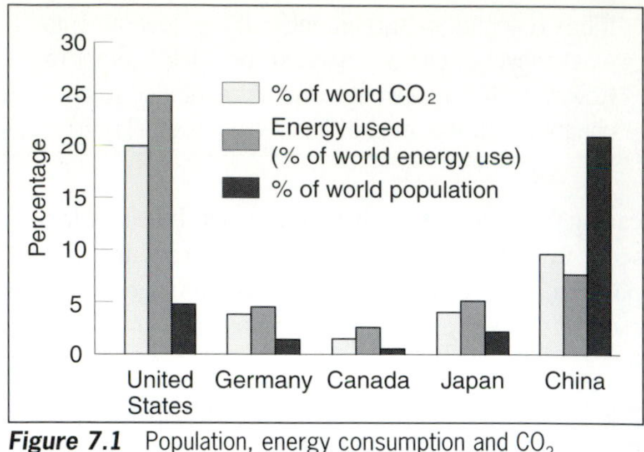

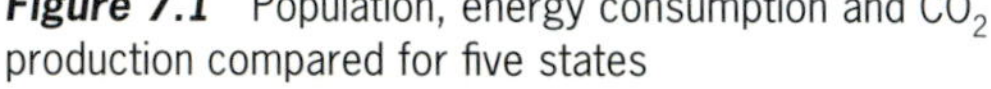

Figure 7.1 Population, energy consumption and CO_2 production compared for five states

When benefits or costs arising from economic activities occur without being captured by prices, the results are termed externalities. Thus, if British Columbia invests in salmon fisheries, but Alaska benefits, the uncompensated losses caused to Canada would be termed an externality. Geopolitical disputes over the environment are often the result of externalities. Hypothetically, if compensation, fair pricing, or international

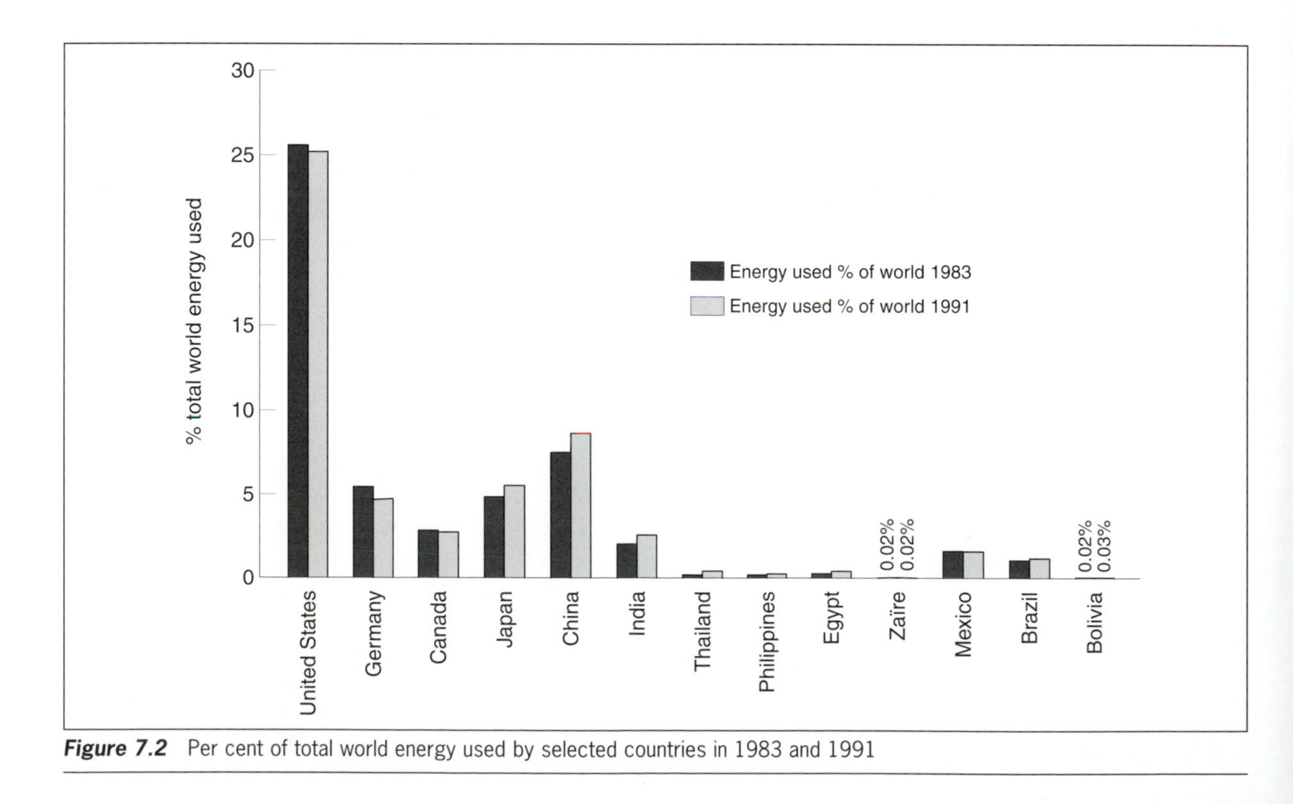

Figure 7.2 Per cent of total world energy used by selected countries in 1983 and 1991

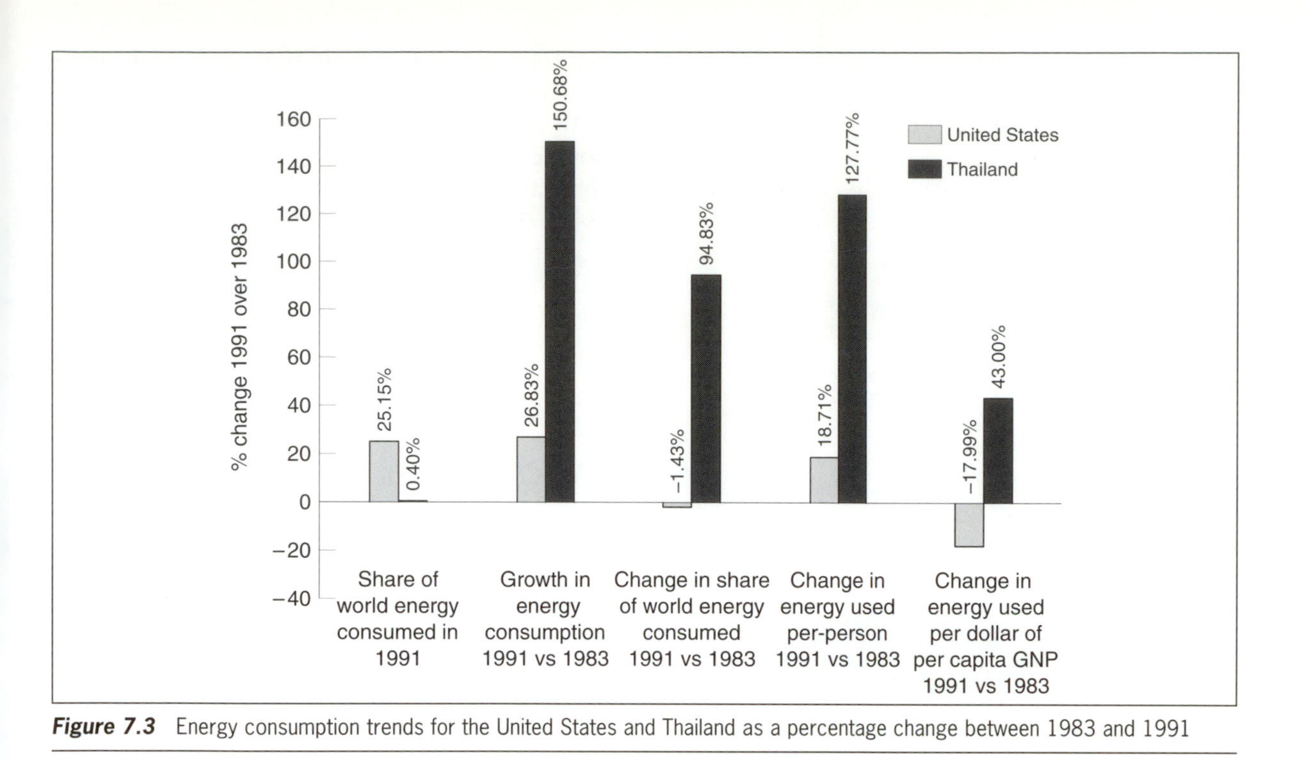

Figure 7.3 Energy consumption trends for the United States and Thailand as a percentage change between 1983 and 1991

agreements could be tapped to eliminate the externalities, many conflicts could be avoided.

7.2.2 CONCEPTS RELATING TO GEOGRAPHY

Externalities are often exacerbated by boundary disputes. The very notion of resource ownership (a Western idea foreign to many other cultures on earth) requires specification of boundaries. Human beings can be remarkably clever about circumventing boundaries that mark off another's rights to a resource, particularly when a fleeting resource is involved (see Figure 7.4).

When ownership (and therefore control and responsibility) over a resource is not specified, the resource may be viewed as belonging to everyone and to no one. Under market economic philosophy, the resource is then common property. It was Garrett Hardin who first referred to "the Tragedy of the Commons" (Hardin 1977). He conceptualized a large pasture owned in common by many herders, each of whom earned a living from herding cattle on the pasture. Knowing that profits increase as more and more cattle are added, each herder has an incentive to increase herd size, and thus cattle begin to overgraze the resource. As the cattle starve, none of the herders gain any profit. The incentive is to use up a resource as quickly as possible before the other stakeholders gain equal access.

Resources such as clean air and fresh water have traditionally fallen into this category, since the idea that any one state or person could own the air seems impossible. Yet, a lack of ownership rights has led to externalities such as acid rain and trans-boundary pollution problems. For example, acid rain generated from the industrial Great Lakes states has caused serious environmental problems in Quebec and the Atlantic Provinces of Canada. Some environmentalists (for example, Peter Berg, with his notion of bioregionalism; see Parsons 1985) argue that taking human relations with the earth back to the local scale may help mitigate against disasters that are of an increasing global scale. The idea is that the further one is removed from close-at-hand impacts, the more one loses sight of responsibilities for wise use or for conservation. In addition, questions of justice and equal access to benefits from resources accompany the need to examine potential geopolitical conflicts. Just as wealthier individual members of a society

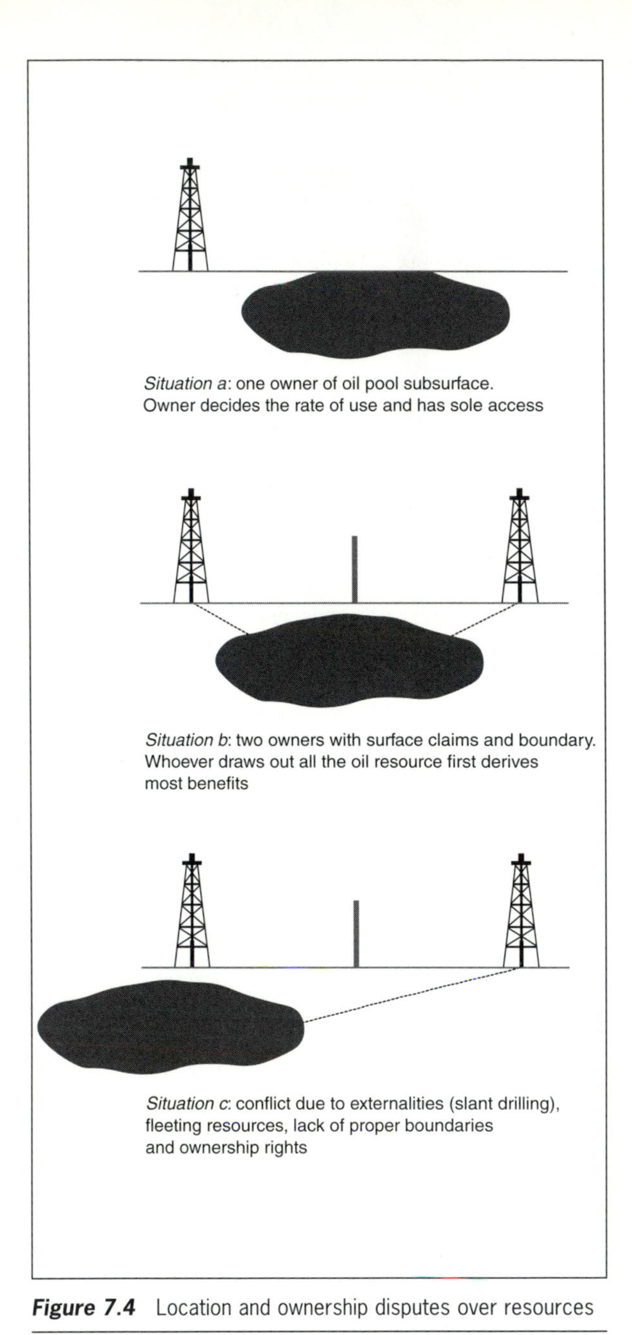

Figure 7.4 Location and ownership disputes over resources

are most protected from the impacts of environmental disaster, and people in poor neighborhoods often have to live with toxic waste dumps or industrial pollution, more developed states may have the ability to export the very environmental havoc they create. Several European states, for example, have negotiated agreements with West African governments to export highly toxic wastes. Therefore, privileged member states can most easily insulate themselves from degradation.

All these economic and geographic concepts of resource management provide useful tools for appreciating the complexity of issues. As seen in Table 7.1, the example of disagreement between the United States and Canada over Pacific salmon can be organized according to the underlying concepts of the dispute.

7.3 Inventory of global environmental issues

The global scale of ecological damage suggests that divisions between wealthy and developing states have contributed to a crisis for the planet's biosphere. The North has created a model for global growth that includes consumption patterns difficult to sustain, and the South appears to be adopting that model. All the issues outlined below have in common a transboundary element to the problems and a demand for interstate cooperation to solve them.

7.3.1 GLOBAL WARMING ("GREENHOUSE EFFECT")

Evidence appears conclusive that human activities are having an impact on the carbon cycle (see Figure 7.5) and the atmospheric temperature of the earth. Burning of fossil fuels (coal, oil, gas) and of wood (use of trees for fuel by a burgeoning population in the developing world, coupled with slash and burn agriculture) are increasing carbon levels. The carbon dioxide compound (CO_2) in turn acts as a blanket to keep the sun's re-radiated energy down near the surface of the earth, thus creating an effect akin to a greenhouse.

The burning, or emissions, or fossil fuels is a worldwide byproduct of modernization and industrialization; therefore, the problem demands global cooperation in finding a solution. While wealthier states still consume a vast share of the world's energy (see Figure 7.6), developing states are experiencing a higher growth rate in fuel consumption (see Figure 7.3). The externality potential of this issue is enormous, since action by a major fuel-

Table 7.1 A fish story revisited (an application of chapter concepts)

Concept	*Example shown by case study*
Resource	Chinook salmon
Renewability	Stocks can replenish if not overdrawn
Fleeting resource	Anadramous fish move between fresh water and open ocean
Transboundary	Resource transcends Canadian and US waters; fishers may cross international boundaries to get to resource
Common property	Cannot specify who owns fish exactly
Rate of use	Not just determined by market; opportunity to harvest is lost if other parties get there first
Externality	Overfishing by one party draws down stock for all
Treaties	1985 US–Canada Treaty; 1996 agreement between states and Indian tribes
Enforcement	Dependent on the goodwill of two sovereign states
Penalties	Canadian tax on boats for transit; threatened trade sanctions
Interest groups	
international	Pacific Salmon Commission
national	Canadian and United States governments
sub-state	Governments of British Columbia, Alaska, Oregon, Washington, Indian tribes
private	Commercial fishers, sports fishers, consumers, scientific and environmental groups
Scarcity of resource	Chinook salmon diminishing to point of non-renewability
Sustainability	Depends on compliance of all parties

BOX 7.3

Resource concepts outside the market

Viewpoints outside the realm of market economics are indicative that dissatisfaction with the dominant model is emerging. The idea that individual consumption, ownership rights, and capitalist structures can best determine utilization of resources has been brought into question. *Ecofeminism*, for example, emphasizes the link between human treatment of the natural world and the role of women in Western cultures, holding that many of our assumptions about nature are based on a patriarchal outlook that often leads to dualism (humans and nature are separate) and dominion as the primary right of people over nature.

Deep ecology, begun through the writing of Arne Naess (1983), and developed into his personal worldview ("Ecosophy T"), attempts to present a new philosophy of humanity and the earth, in which humans identify self more with totality of the earth ecosystem. Values are intrinsic in nature, and do not need to be realized through the marketplace. While such new movements are emerging, coupled with an interest in non-Western religious traditions about the environment, some argue that resource use is still best realized through market systems, as the obviously poor record of centrally planned economies suggest. Whatever viewpoints prove most socially acceptable over the long run, the increasingly transboundary and global nature of these relations suggests that environmental questions will continue to occupy center stage in geopolitics into the twenty-first century.

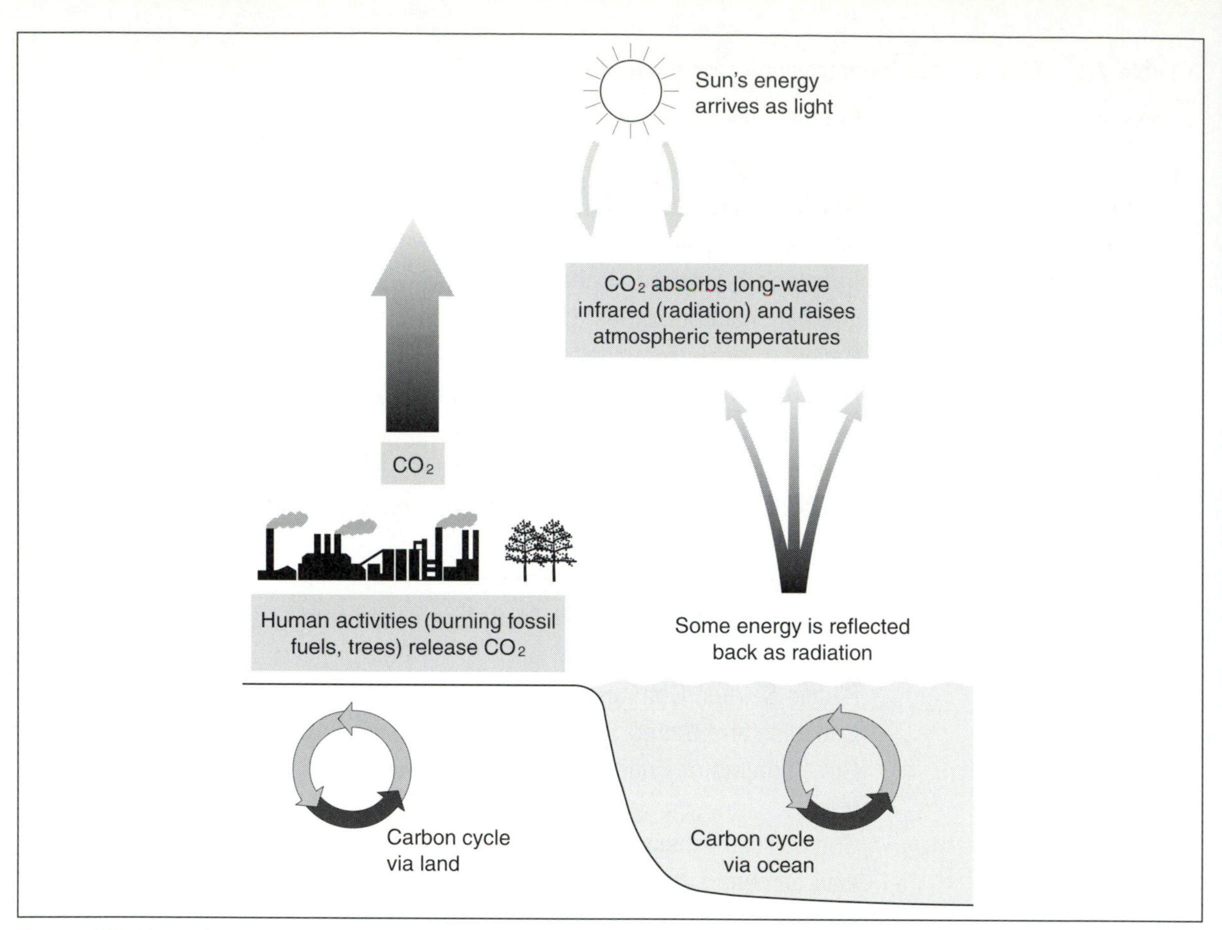

Figure 7.5 The carbon cycle is a process that makes carbon available to all life on earth. Carbon is a non-metallic chemical element, the sixth most abundant element in the universe. Carbon dioxide is a compound of carbon that makes up 0.05% of the atmosphere. When plants and animals decompose under certain conditions, carbon is tied up, to be released back into the cycle when humans burn it (as fossil fuels such as coal, oil, and gas). Burning trees also releases carbon. The ocean serves as a great repository for carbon through the food chain, through dissolved ocean water, and through bottom sediments

Plate 7.1 Carbon emissions from rush-hour traffic on a freeway, Seattle, WA (photo: Kathleen Braden)

using state can have a profound impact on the outcome for all parties. China, for example, is heavily reliant on coal for fuel (see Figure 7.2), and the demand for coal in China is expected to double over the next 25 years as the economy continues to expand. In addition, coal is burned inefficiently in China, and technology to scrub or clean emissions (especially from power plants) is costly for a state of over one billion people. In addition to China, the United States and the former USSR, Japan, Germany, India, and Canada are all major producers of CO_2 (see Figure 7.7).

Scientists forecast an increase in average global temperatures of 1.44–6.3°F (0.8–3.5°C) by the year 2100 at current emission levels of greenhouse

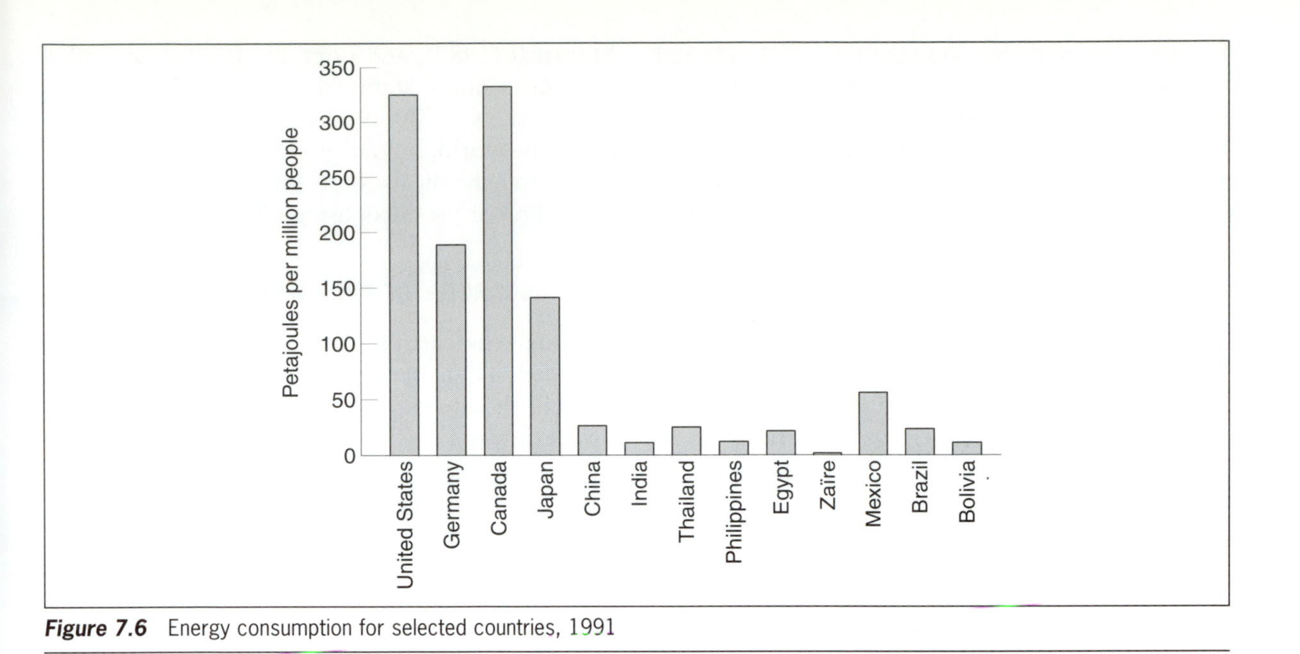

Figure 7.6 Energy consumption for selected countries, 1991

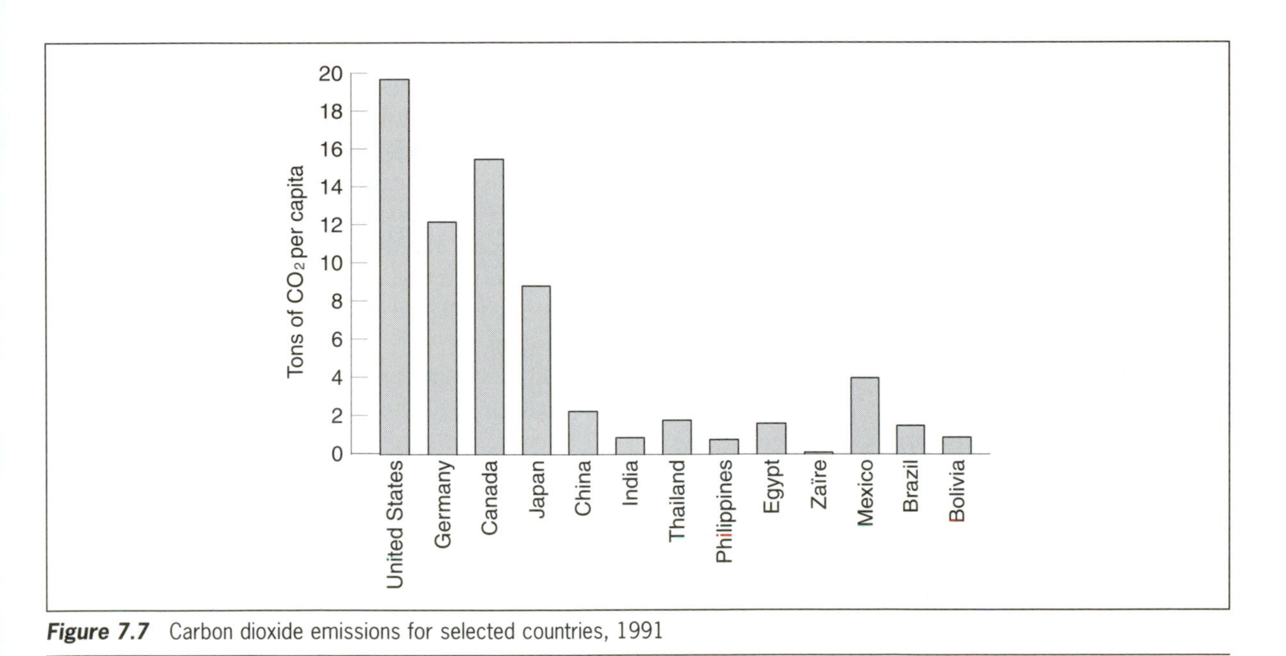

Figure 7.7 Carbon dioxide emissions for selected countries, 1991

gases (carbon dioxide, methane, nitrous oxide). Arguments within the scientific community over the efficiency of computer models designed to simulate the impact of greenhouse gases may have delayed attempts to deal with the problem. The likely impact will probably be harmful to agriculture, create havoc with weather systems, and perhaps speed up melting of polar ice, raising the sea level. Governments of states in the Pacific would be most severely affected by a rise in ocean levels, since many (such as Kiribati) are located on low-lying lagoons where supplies of fresh water could be the first to be impacted by salt saturation.

At the 1992 United Nations Conference on

Environment and Development (UNCED, referred to popularly as the Earth Summit) in Rio de Janeiro, more than 150 states signed a framework for dealing with greenhouse emissions. The treaty went into effect in 1994, having achieved the required number of ratifications by national legislatures. However, fulfilling the agreement calls for development of measurement, control, and verification tools.

In December 1997, 160 states reached an agreement in Kyoto, Japan, for limiting greenhouse gas emissions. On average, industrialized states are obliged to reduce their emissions by an average of 5 per cent below 1990 levels.

7.3.2 DEPLETION OF THE EARTH'S OZONE

A second major environmental concern with transboundary implications is the loss of protective ozone (a form of oxygen) from the earth's atmosphere. It tends to accumulate in a layer approximately 15 miles (24 km) above the earth, but its quantity varies seasonally. Ozone protects the earth from excessive sun energy because it absorbs ultraviolet light.

Scientists studying the depth of the ozone layer first noticed that it was diminishing over the southern polar regions entirely for a few months each Antarctic summer. Evidence is now available that production of CFCs (chlorofluorocarbons) used widely in industry for items such as insulation (refrigerators), were interacting with ozone and depleting it, particularly in the Southern Hemisphere over Antarctica. Countries adjacent to or encompassed by this "hole in the ozone layer", such as Chile, have become increasingly alarmed and warned their citizens about sun exposure. In 1995, the hole expanded at an unprecedented rate and covered an area twice the size of Europe. In the Northern Hemisphere, increasing concentrations of chlorine monoxide over the Arctic area in the early 1990s propelled European states to push for a faster implementation of the ban on ozone-depleting chemicals. The United States, which alone of the seven largest industrialized states (Group of Seven) originally wanted to slow down the elimination of the chemicals, agreed in 1992 to a faster paced phase-out.

International agreements in Vienna (1985), Montreal (1987), and London (1990) have called for industrialized states to cut CFC production by half by the year 2000. Output of CFCs is declining in the world, but developing states still find the use of the chemicals appealing, and international smuggling of products using CFCs has occurred.

7.3.3 SCARCITY OF FRESH WATER

Although the earth's water supply has long been regarded as a renewable resource, it too has become threatened at the global level. Demands for fresh water due to population growth, expansion of irrigated agriculture, dam construction, water pollution, and consumption by industry have meant water shortages in more than 40 per cent of the world's countries. Demand for water is increasing at approximately 2.3 per cent per year. The World Bank, recognizing the crisis, is spending $600 billion before the year 2005 to increase supplies of fresh water. Shortages are especially severe in countries with arid environments, but the big consuming states tend to be the industrialized ones (see Figure 7.8). Consumption of contaminated water leads to diarrhea and chronic diseases, especially among children, and is one of the primary killers of young people in the developing world; almost one billion people in the world's rural areas may be consuming contaminated water (World Resources Institute 1994).

Sovereignty over water resources is a difficult geopolitical issue to solve. Moving as a fleeting resource through the water table or on the surface via interstate rivers, water is a trespasser of human boundaries. The actions of upstream states can have severe externality impacts on users downstream. Most major rivers in the world have been subject to some degree of disagreement by the states using them (see Figure 7.9). The River Rhine in Europe flows through Switzerland, Germany, and the Netherlands, with ongoing industrial pollution along the length of its course. The states of Central Asia in the former USSR pollute or withdraw for irrigation water from the Amu Darya and Syr Darya rivers that once fed into the Aral Sea. India and Bangladesh share ownership rights to the Brahmaputra and Ganges river systems, leading to conflicts between the states. Similarly, India and Pakistan have long dis-

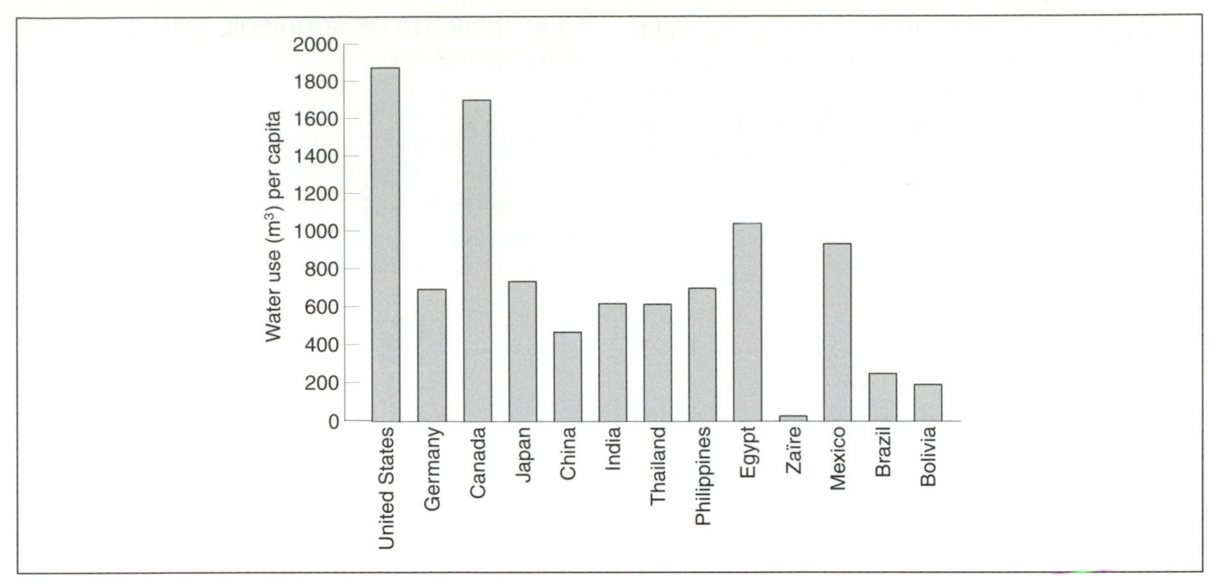

Figure 7.8 Fresh water withdrawals per capita, various years

Figure 7.9 Selected international disputes over pollution or withdrawals of river water

puted the withdrawal of water from the Indus River. Perhaps the most explosive situation exists in the Middle East, along the Jordan River valley, where water withdrawals for irrigation and population needs have increased pressure on the river by all states in the region.

7.3.4 SCARCITY OF GOOD SOIL AND DESERTIFICATION

The world contains approximately 13.4 billion hectares of land, but not all of it contains soil that is useful for human economic activities. Approximately 22 per cent of the earth's surface can produce crops without irrigation water, but fertile soil is being used up rapidly. The United

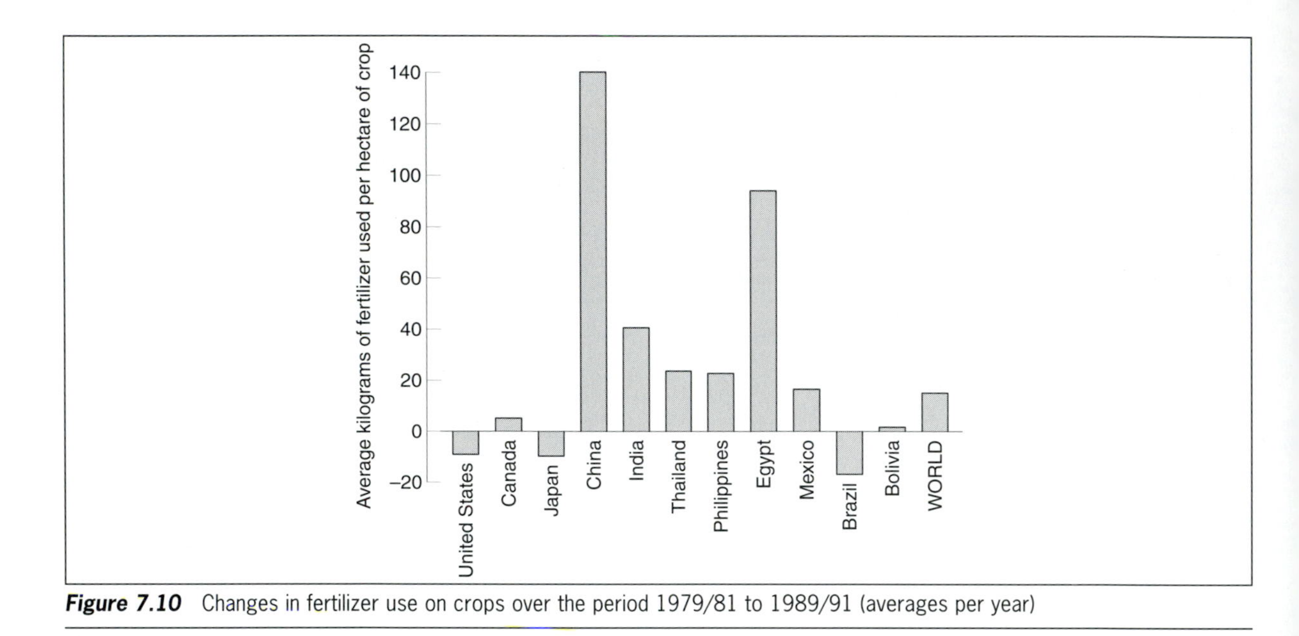

Figure 7.10 Changes in fertilizer use on crops over the period 1979/81 to 1989/91 (averages per year)

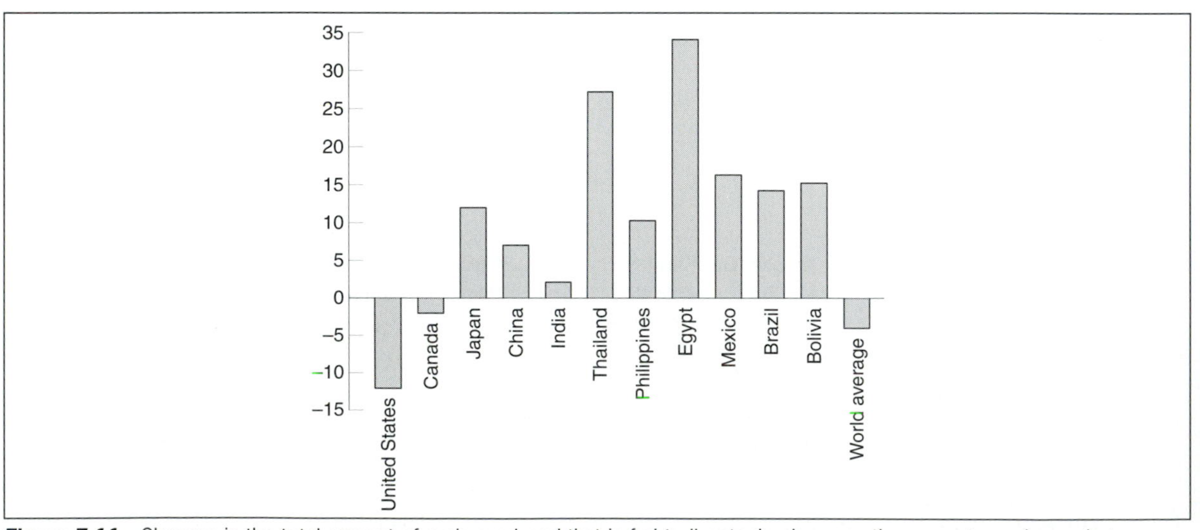

Figure 7.11 Changes in the total amount of grain produced that is fed to livestock, shown as the percentage change between 1972 and 1992

Nations Environment Program (UNEP) Global Assessment of Soil Degradation study found that 10–11 per cent of the earth's fertile soil has been damaged to the point that it may be impossible to reclaim. Two-thirds of the soil seriously eroded is in the developing countries of Asia and Africa, but Central America has also experienced the loss of about 25 per cent of its good soil. The main causes have been poor agricultural practices and overgrazing by livestock. Loss of productivity is in part being compensated by the increased use of fertilizers and higher-technology methods, but such measures are costly in developing economies. Despite the cost, the trend in recent decades has been for the rapid growth of fertilizers (see Figure 7.10) and chemical pesticides in developing states.

The fertile soil remaining is increasingly being used to produce crops that are fed to livestock, rather than for direct human consumption. As noted in Figure 7.11, from 1972 to 1992, developing states such as Egypt devoted an increasing share of their grain production to livestock as traditional systems gave way to greater demands for meat.

The spatial spread of deserts has also removed land from utilization, particularly for livestock. In 1977, the United Nations held a conference on desertification in Nairobi, Kenya, and defined desertification as the impoverishment of terrestrial ecosystems, which can be measured by reduced productivity of desirable plants, alteration in the biomass, and accelerated soil erosion. The United Nations estimated that 27 million hectares of land are removed from agricultural use yearly through the process, and that one-third of the earth's land is threatened; 80 per cent of arid rangelands. Sub-Saharan Africa has felt the most severe impacts thus far. On the African Horn, the Inter-governmental Authority on Drought and Development, which includes Ethiopia, Djibouti, Kenya, Sudan, Somalia, and Uganda, was formed to deal with desertification and the resulting displacement of people. The organization has also attempted to settle nomadic people of the region, but competition for resources, traditional lifestyles, and wars in the region have made the effort difficult.

7.3.5 AIR POLLUTION, ACID RAIN

One form of air pollution with geopolitical consequences has been acid rain. Sulfur dioxide and nitrogen oxide from fossil fuel burning combine with water vapor to create sulfuric acid and nitric acid, which is then airborne across boundaries. Forests and water bodies particularly suffer when the acidic precipitation is deposited. Along with eastern North America and eastern China, Europe has experienced challenges associated with acid rain; 25 per cent of Europe's forests have been damaged. Dirty industrialization in Central Europe has been a problem for the Scandinavian states as acid rain is carried north. As a result, 34 European governments have signed the Long Range Transboundary Air Pollution Convention in an attempt to reduce emissions. The wealthier Scandinavian states have initiated programs of technical assistance to states formerly in the Soviet bloc to help reduce pollution that may impose externalities on Scandinavian populations.

7.3.6 TOXIC WASTE DISPOSAL

Industrial activities generate large amounts of waste which often add toxicity to the environment when dumped. Toxic metals include lead, zinc, copper, arsenic, antimony, and mercury, all of which can be harmful to life forms. Mining processes are especially large producers of metal wastes, and mine tailings (byproducts from mine excavations) are merely dumped in place on the landscape in many states. Contamination of the water table can then have externality impacts on downstream states. Chemical wastes and toxic air emissions, nuclear wastes, and toxic compounds from fertilizers are also prime threats to the environment.

Firms in India, for example, often discharge waste directly into streams or dump waste onto land without mitigation measures. All of the major rivers in eastern China have been polluted with industrial discharge; mining operations in states such as Brazil and Kyrgyzstan release heavy metals such as mercury into the groundwater.

The transfer of noxious wastes has particular geographic implications as states may not always be aware that shipments through their boundaries may contain harmful byproducts. In addition, wealthier states where environmental regulations may be more stringent are finding it in their

BOX 7.4

Going nuclear: the ultimate transboundary toxic waste

Climate models suggest that a massive exchange of nuclear weapons (e.g. between Russia and the United States) could place enough dust and particles in the atmosphere to lower the world's temperature and create a "nuclear winter" effect. The result of even a few degrees temperature drop in the Northern Hemisphere would cause rice crop failures and the death of billions from starvation. The Chernobyl nuclear power accident in 1986 contaminated not only the Ukraine where it originated, but nearby Belorus and much of northern Europe as well. Disposal of nuclear waste from power plants and other civilian uses, as well as the accumulation of weapons-grade plutonium (and increased evidence that the material is being smuggled internationally), suggest that development of nuclear capabilities in war and peace have the potential to create massive externalities on the world's states.

interest to ship toxic wastes to poorer countries for disposal. The United Nations Convention on Control of Transboundary Movement of Hazardous Wastes and Their Disposal has run into controversy during the ratification process over what exactly constitutes hazardous waste versus "clean" recycling or scrap metals. Industrialized states such as Germany have expressed reservations, and the United States has declined to ratify the convention. Accusations were traded between China and the United States over shipments of radioactive and medical wastes from the US, planned for disposal in China.

Finally, what of waste disposal in regions not covered by state boundaries? Dumping of wastes at sea is continuing on the planet, usually on an illegal basis. Radioactive waste and decommissioned weapons systems are often disposed of at sea by states, creating externalities for coastal or fishing countries.

7.3.7 DEFORESTATION AND LOSS OF BIODIVERSITY

Trees are considered a renewable resource, yet throughout human history, forests have been cut down to the point of extinction. Many of the forests of the Northern Hemisphere (in Europe and North America) have already been utilized or transformed into plantation crops of trees. For example, virtually all of what is now the eastern United States was forested at the time of Columbus, yet by 1900, much of the original forest cover was removed. Likewise, the region the British call the moors was forested in recent human history. In fact, by the year 1915, 75 per cent of all the deforestation that has occurred on the globe had already taken place.

Harvesting of tropical wood has been proceeding aggressively over the past 40 years. The United Nations Food and Agricultural Organization estimates that from 1980 to 1990 alone, forest area declined by 7 per cent in Central Africa, 6.6–13 per cent in Southeast Asia, 7 per cent in South America, and 21 per cent in Central America.

Wealthier states consume more wood products per capita, but demand has been rising most quickly in the developing world. Wood is required for commercial use (roundwood consumption has doubled during the last 30 years) and for fuel in countries such as Nepal and Haiti, which have few affordable alternatives.

But does this rapid draw down of tree stocks have geopolitical implications, since trees are not a fleeting resource? Transboundary effects of deforestation may be felt in three ways:

1 erosion of soil and impact on downstream states as protective tree cover is removed;
2 reduction of CO_2 absorbing capacity and contribution to the global warming phenomenon;
3 loss of biodiversity.

Forests are biological communities of plants and fauna, and may be one of the most important biomes for preserving diversity of earth life forms. In recognition of this importance, a convention on preserving the earth's biodiversity was a hallmark of the 1992 UNCED process. The United Nations created a Commission on Sustainable Development, which in turn established a Panel on Forests

Plate 7.2 Snow leopard skins are still being traded on black markets in Asia despite the CITES agreement. Source: International Snow Leopard Trust

under the leadership of Canada and Malaysia (two major producing states of industrial roundwood).

Biodiversity is defined as all species of plants, animals, and microorganisms, and all the ecosystems and ecological processes of which they are a part. In one sense, all of the global ecological problems created by people may be considered to come under the heading of threats to biodiversity. We may be accomplishing classification of astronomical features faster than of terrestrial life. The extinction of plants and animal species thus results in a loss of biodiversity.

In the past 300 years, 1,622 extinctions of species have been documented, and more than 26,000 are currently considered threatened. The rate of species loss may be 1,000 times higher than the loss rate would be without human agency. However, measurements of extinction rates are very controversial and imprecise.

Extinction, however, is also a natural process, and societies may vary greatly in their value systems. Does conservation of biodiversity resemble the issue of global warming with its implications at the global scale? Perhaps biodiversity maintenance is of local value only.

Four ideas suggest that biodiversity loss can be a catastrophe of global dimension:

1. Agricultural products are derived from a diversity of wild species. Foods such as oranges, coffee, cocoa and many others originated in tropical forests, and the pharmaceutical industry constantly derives new medicines from species found in forests. Because simple ecosystems are less stable than complex ones, a planet earth in the future, totally modified by human agency, will be less resilient against diseases and defects than a managed and simplified set of ecosystems.
2. Species that migrate across international borders meet the definition of fleeting resources, and international conventions for their protection are in place, although not always honored. Impacts of one country's actions on the food chain may not always be discernible on the biodiversity of a downstream state until it is too late.
3. Even when the value of species cannot be readily measured (unlike the rare case when profit can be calculated from a burgeoning ecotourism sector), there may be spiritual or psychological values that are real, if difficult to price. Such non-priced values can be global in character.
4. Interruption of biodiversity maintenance at the biome level may have biosphere implications, since interactions at a planetary scale are only now becoming clear. For example, recent studies are indicating that the health of the Amazon rainforest may be linked to that of the drylands of northern Africa, with tons of dust blown westward from Africa every year and deposited in Latin America.

Ever more sophisticated modeling techniques are revealing that few environmental disruptions are spatially limited. From the greenhouse effect to the extinction of species, environmental security and international politics seem increasingly linked.

7.4 The interest groups

Since impacts of environmental damage are increasingly at the global level, supranational bodies play a greater role in their resolution. This move to the international arena, however, comes at the price of threats to international sovereignty and ownership rights over resources. Externality disputes can rarely be solved state-to-state without some form of assistance or intervention from

BOX 7.5

Transboundary nature reserves

Conservation of biodiversity is best accomplished through an integrative approach, rather than focusing on single-species management. Reserved lands, or protected areas, can be set aside from most forms of economic use to accomplish the task, but sometimes ecosystems cannot be protected within the borders of just one state. The International Union for the Conservation of Nature and Natural Resources keeps an inventory of the world's protected areas. IUCN notes that much of the Mount Everest ecosystem in the Himalayan Mountains is protected thanks to the cooperation of two neighbors. On the Chinese side, there is the Qomolangma Nature Preserve (3.5 million ha), while Nepal has set aside Langtang (171,000 ha), Sagarmatha (115,000 ha), and Makalu-Barun (233,000 ha) National Parks (Figure 1). The Himalayan species that inhabit the parks can go about their business and they do not even need passports.

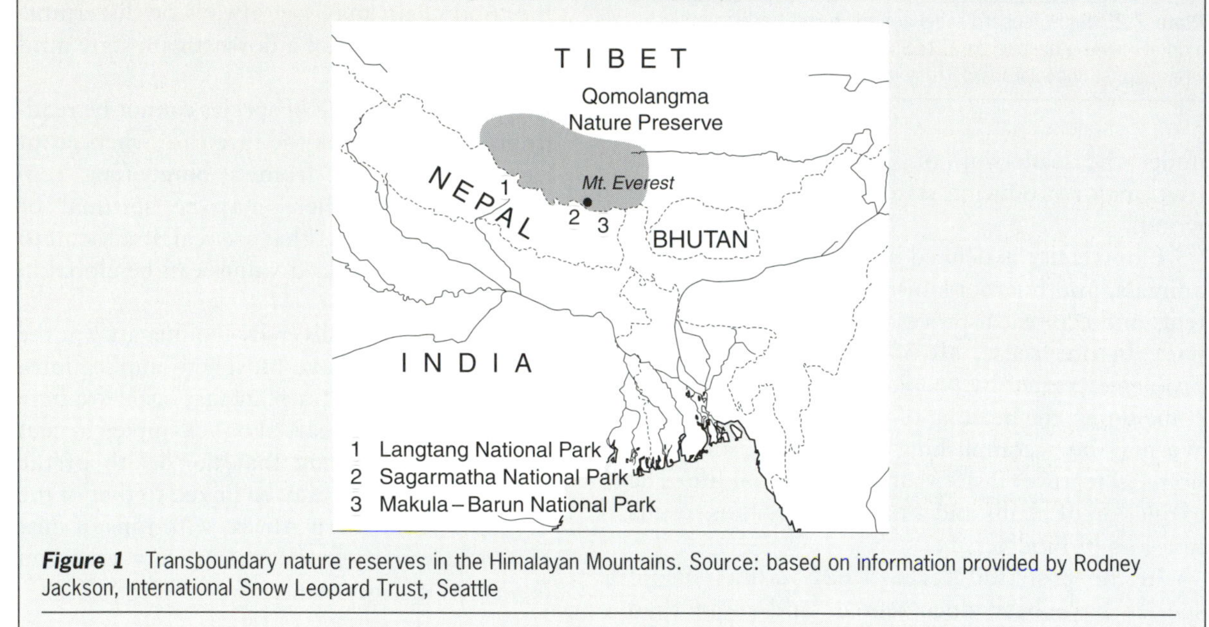

Figure 1 Transboundary nature reserves in the Himalayan Mountains. Source: based on information provided by Rodney Jackson, International Snow Leopard Trust, Seattle

supranational organizations. Thus, the United Nations, the World Bank, and other multinational interest groups recognize the emergence of international environmental conflicts (termed "IEC" in a 1992 UN study cited by Jon Martin Trolldalen) as making up a separate category of interstate issues. This trend in supra-statism has been offset by a counter argument that says environmental problems are best solved at the sub-state (local) level, where impacts are most closely felt.

7.4.1 SUPRANATIONAL ACTORS

The United Nations has become the primary vehicle for resolution of state-to-state conflicts over resources and environmental degradation, often accompanied by concerns for economic development by the South. In 1972 the UN Conference on the Human Environment, held in Stockholm, recognized that protection of the environment could best be done through emerging international law, but that sovereignty questions would impede the process.

Within the United Nations, the UN Environmental Program (UNEP) is an ongoing program of the General Assembly, and especially deals with monitoring environmental issues, providing reports and plans, and dealing with legal instruments. The UN Development Program (UNDP) has also become a major actor as it

Table 7.2 Examples of multilateral agreements to protect the environment

Agreement	*Year signed or in effect*
Hague Declaration Concerning Asphyxiating Gases	1899
Treaty Relating to Boundary Waters and Questions Arising Between the US and Canada	1909
Convention between the United States and Other Powers Providing for the Preservation and Protection of Fur Seals	1911
Protocol for the Prohibition of the Use in War of Asphyxiating, Poisonous or Other Gases and of Bacteriological Methods of Warfare	1925
Convention on Nature Protection and Wildlife Preservation in the Western Hemisphere	1940
International Convention for the Regulation of Whaling	1946
International Convention for the Protection of Birds (Paris)	1950
International Convention on North Atlantic Fisheries	1950
Oslo Agreement on Protection of Prawns, Lobsters, and Crabs	1952
North Pacific Fisheries Act (Japan, US, Canada)	1954
Convention for the Prevention of Pollution of the Sea by Oil	1954
International Agreement on Protection of Fur Seals	1957
Antarctic Treaty	1959
Paris Convention on Third Party Liability in the Field of Nuclear Energy	1960, 1963
Declaration on the Prohibition of the Use of Nuclear and Thermo-nuclear Weapons (UN Resolution 1653)	1961
Treating Banning Nuclear Weapons Tests in the Atmosphere, in Outer Space, and Under Water	1963
Vienna Convention on Civil Liability for Nuclear Damage	1963
Algiers Convention on Conservation of Nature and Natural Resources in Africa	1968
International Convention on Civil Liability for Oil Pollution Damage	1969
Benelux Convention on the Hunting and Protection of Birds	1970
Brussels Convention Relating Civil Liability in the Field of Maritime Carriage of Nuclear Material	1971
International Convention on the Establishment of an International Fund for Compensation of Oil Pollution Damage	1971
Ramsar Convention on Wetlands of International Importance Especially as Waterfowl Habitat	1971
Convention for the Conservation of Antarctic Seals	1972
London Convention on the Prevention of Marine Pollution by Dumping Wastes and Other Matter	1972
Oslo Convention for the Prevention of Marine Pollution by Dumping from Ships and Aircraft	1972
Convention for the Protection of the World Cultural and Natural Heritage	1972
Convention on the Prohibition of the Development, Production, and Stockpiling of Bacteriological (Biological) and Toxin Weapons and on Their Destruction	1972
Convention for the Prevention of Pollution by Ships (MARPOL)	1973
Convention on International Trade in Endangered Species of Wild Fauna and Flora (CITES)	1973
International Agreement on Protection of Polar Bears	1973
Nordic Environmental Protection Convention	1974

Table 7.2 Continued

Agreement	*Year signed or in effect*
Helsinki Convention on the Protection of the Marine Environment of the Baltic Sea Area	1974
Paris Convention for the Prevention of Marine Pollution from Land-Based Sources	1974
Barcelona Convention for the Protection of the Mediterranean Sea Against Pollution	1976
Convention on the Prohibition of Military and Any Other Hostile Use of Environmental Modification Techniques (UN Resolution 31/72)	1976
Berne Convention on the Conservation of European Wildlife and Natural Habitat	1976
London Convention on Civil Liability for Oil Pollution Damage Resulting from Exploration for and Exploitation of Seabed Mineral Resources	1977
Kuwait Regional Convention for Cooperation on the Protection of the Marine Environment from Pollution	1978
Geneva Convention on Long-Range Transboundary Air Pollution	1979
Bonn Convention on the Conservation of Migratory Species of Wild Animals	1979
International Agreement on Protection of Vicuna	1979
Convention on the Conservation of Antarctic Marine Living Resources	1980
Lima Convention for the Protection of the Marine Environment and Coastal Area of the South-East Pacific	1981
United Nations Convention on Law of the Sea (UNCLOS)	1982
Jiddah Regional Convention for the Conservation of the Red Sea and Gulf of Aden Environment	1982
Cartagena Convention for the Protection and Development of the Marine Environment of the Wider Caribbean Region	1983
Vienna Convention for the Protection of the Ozone Layer	1985
Nairobi Convention for the Protection, Management, and Development of the Marine and Coastal Environment of the Eastern Africa Region	1985
US–Canada Pacific Salmon treaty	1985
Convention on Early Notification of a Nuclear Accident	1986
Convention on Assistance in the Case of a Nuclear Accident or Radiological Emergency	1986
Noumea Convention for the Protection of the Natural Resources and Environment of the South Pacific Region	1986
Protocol on the Reduction of Sulphur Emissions or Their Transboundary Fluxes by at Least 30 Per Cent 1985	1988
Protocol Concerning the Control of Emissions of Nitrogen Oxides	1988
Wellington Convention on the Regulation of Antarctic Mineral Resources	1988
Basel Convention on the Control of Transboundary Movements of Hazardous Wastes and Their Disposal	1989
Protocol on Reduction of Volatile Organic Compounds	1991
Bamako Convention on the Ban of Import into Africa and the Control of Transboundary Movement and Management of Hazardous Wastes within Africa	1991
Espoo Convention on Environmental Impact Assessment in a Transboundary Context	1991
Protocol on Environmental Protection, Antarctica	1991

Table 7.2 Continued

Agreement	*Year signed or in effect*
Convention on Biological Diversity	1992
Convention on Climate Change	1992
Convention on the Prohibition of the Development, Production, Stockpiling and Use of Chemical Weapons and on Their Destruction	1993
United Nations Resolution Against Drift-Net Fishing	1993
Montreal Protocol on Substances that Deplete the Ozone Layer	1987, 1990

attempts to influence the economic development direction of member states. Associated groups of the UN, such as the FAO (Food and Agricultural Organization) and the World Health Organization (WHO) also play important monitoring and program roles.

UNCED resulted in various international conventions and commissions. However, before this conference, international regulations on environmental damage had been well under way both at the UN and at regional levels, extending back to the nineteenth century in the case of noxious gases (see Table 7.2 on p. 115–117).

Emerging as a major player, particularly since the Earth Summit, the World Bank (formally, the IBRD or International Bank for Reconstruction and Development) has allocated a growing amount of time, attention, and funds to environmental concerns. In earlier decades, the Bank came under heavy criticism from environmentalists for financing projects that often had detrimental impacts on local ecosystems and failed over the long run to provide promised economic benefits. UNCED established a Commission on Sustainable Development (to monitor compliance with the various conventions on the environment) and further strengthened the GEF, or Global Environmental Facility. The World Bank, UNEP, and UNDP had set up the GEF in 1991 to provide a funding mechanism for carrying out the programs that emerged after the Earth Summit.

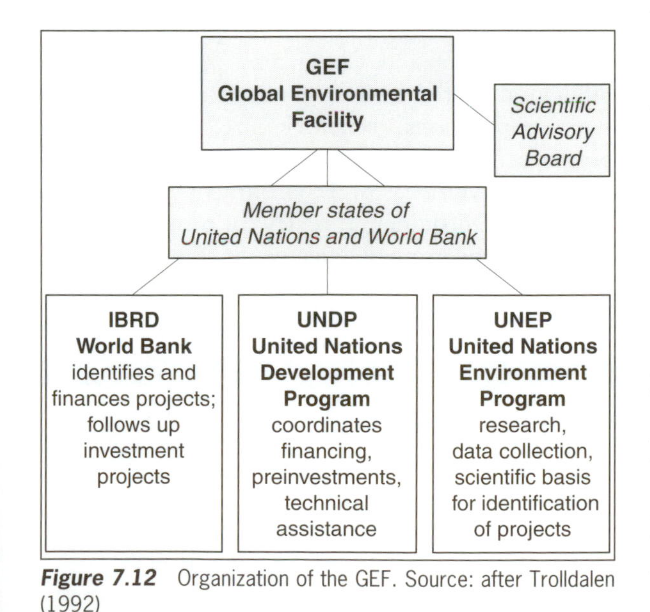

Figure 7.12 Organization of the GEF. Source: after Trolldalen (1992)

The GEF is designed to tackle four issues: global warming, pollution of international waters, destruction of the ozone layer, and preservation of biodiversity. The idea is that work in host countries will have positive spillover benefits globally. The first phase of GEF included about $1.2 billion in financing, mainly earmarked into a Global Environmental Trust Fund. Twenty-eight states of the industrialized world pledged money. The GEF is undergoing several phases of restructuring, but includes the basic organization noted in Figure 7.12.

Regional organizations of sovereign states have also played a role in environmental disputes. Two examples are the OAU (Organization for African Unity), which passed a resolution against the dumping of toxic waste in Africa, and the Conference on Security and Cooperation in Europe, which discussed emissions controls for air pollutants.

However, international regulation of externalities is hampered by the fact that sovereignty rights

BOX 7.6

Bhopal: the role of multinational firms

On December 3, 1984, one of the worst industrial accidents the world has witnessed occurred as toxic methyl isocyanate gas leaked from a Union Carbide pesticide plant in Bhopal, India. The initial death count was recorded at 2,500, but estimates in 1997 of subsequent deaths show that the toll may have reached 10,000. The difficulty in providing compensation to victims has shown the problems associated with liability for environmental damage. Union Carbide had the case tried in Indian court, paid a relatively small settlement, and even those funds were held up by the Indian government for eight years.

of states still give individual governments room to be non-compliant. Monitoring and enforcement usually depends on the goodwill of participating states. At the forefront of international environmental law now are issues of sanctions, compensation, and international taxing authority – all measures creating great controversy.

Other actors at the supranational level include large international organizations designed to tackle environmental questions, such as the International Union for Conservation of Nature and Natural Resources (IUCN), based in Switzerland. IUCN, founded in 1948 to bring together non-governmental organizations and governments around issues of resource use, now works in over 130 countries. Another international organization that performs a crucial information gathering and analysis function is WRI, the World Resources Institute, which publishes an invaluable annual guide to the world global environment entitled, *World Resources*.

Supranational organizations that influence world trade, such as the General Agreement on Trade and Tariffs (GATT – now the World Trade Organization), the UN Conference on Trade and Development, and international or regional trading blocs, can all have an impact on the environment through their influence on consumption and production patterns. Multinational firms can also be quite influential on the world environment for better or worse. Past controversies over subjects as diverse as the purchase of beef fed with crops requiring tropical deforestation or the manufacture of CFCs in refrigeration demonstrated the impact of policies of multinational firms.

7.4.2 NATIONAL GOVERNMENTS

Sovereign states still form the building blocks of geopolitical relations, and their choices may yet have the greatest impact on the global environment. In 1972, only 25 countries had environmental agencies at the national level, but as of 1995, such agencies existed in over 140 states.

Governments of states must respond to local pressures for economic development and balance these with increasing awareness that environmental destruction will harm development over the long run. India and China may prove especially influential on the global environment due to their large populations, territory size, and growing pressures for economic development. Within India, the Ministry of Environment and Forests regulates environmental review of projects, and in 1986 the government enacted an Environmental Protection Act at the national level. But compliance has been difficult amid a population growth

BOX 7.7

South China Sea and environmental conflict

The Spratlys island chain occupies almost 400,000 square miles (around 1 million km^2) in the South China Sea and might be of little significance to people except for the potential of oil development, fishing, and use for ocean transportation. China, the Philippines, Taiwan, Vietnam, Malaysia, and Brunei all have claims to the territory, and all except Brunei have provided military presence there. Minor clashes over sovereignty of the region have erupted, most recently between the Philippines and China and between Taiwan and Vietnam.

that will take India over the billion person mark in the near future, combined with growing urbanization and industrialization.

China's influence at the global level is perhaps even stronger than India's due to the rapid pace of China's economic growth. Rising incomes, especially in eastern China, have pushed consumption demands for consumer products, and despoliation of urban environments has been especially severe. China's actions also influence its neighbors. Acid rain, pollution of rivers, demand for wildlife organs in traditional Chinese medicine, and a search for energy sources to develop (such as potential oil development in the South China Sea) has placed China at odds with other states in the region. Yet the Chinese government has also been a signatory to important international conventions, such as the one designed to cut CFC production, and is increasing national spending on environmental clean-up.

7.4.3 NON-GOVERNMENTAL ORGANIZATIONS AND POLITICAL MOVEMENTS

Public participation in environmental issues is often fulfilled not only through government structures, but also via non-governmental organizations (NGOs) and increasingly through political processes. Public interest and participation seem to be highest in wealthier states, perhaps because knowledge about power structures is more widespread and demands of daily life less stringent. However, citizen participation is growing in developing states as well; for example, a non-governmental group in China, Friends of Nature, formed in 1994 and has been active in influencing clean-up of the Yangtze River, and environmental activism has long been evident in India.

Some NGOs have taken on international stature, which allows them to become significant actors in world environmental issues: WWF (Worldwide Fund for Nature), Greenpeace, Nature Conservancy, and Earth Island Institute all have programs in a variety of countries. Their philosophies may vary, but they provide a means for citizens to influence geopolitical outcomes separately from governmental channels.

One group, Conservation International, based in the United States, has entered directly into the North–South issue by purchasing debts of developing states from banks. The organization then negotiates an agreement with the developing state in a type of "debt-for-nature" swap. This strategy has also been used on a larger scale by some programs of the IBRD.

The emergence of "green" political parties has also created a new branch of political life. Much of the early impetus for reform in the USSR was centered around discontent with the environmental record of the state, and the green political parties there were the early bell-wethers for political change. Starting in the late 1980s, elections in Europe saw the increasing strength of the Green parties, particularly in Germany and the Netherlands.

Environmental issues have been linked to other social causes or philosophical positions, such as feminism, racial and economic justice, and religion. The role of women, in particular, in moving the earth's economies towards a more equitable and less environmentally harmful stance has been given recognition by the United Nations and World Resources Institute. Preliminary meetings leading up to UNCED included conferences on the role of women, and UNEP includes a Senior Women's Advisory Group. In 1991, a meeting organized by the World Women's Congress for a Healthy Planet numbered 1,500 participants.

The multitude of interest groups that now influence global use of and attitude towards resources demonstrates that we have entered a period of potential increase in environmental conflicts, but growing diversity of views about how to deal with the issues.

On the one hand, the Western market approach to resource conservation is codified through the growing role for international financial mechanisms such as the GEF. Capitalist models for development have become less questioned since the failure of the centrally planned Soviet-style economies; on the other hand, more opportunities for different approaches to the relationship between environmental health and economic development are available. Questions such as: what is development exactly? Can and should current models of consumption and growth be extended spatially across more and more states and across time into an uncertain future sustainability?

Interest groups as well seem to be tending into two apparently contradictory directions. Supranational organizations play an increasing role in influencing states' behaviors with respect to the environment. Yet at the same time, individual citizens and organizations have growing possibilities to be players in global environmental issues. Meanwhile, tugged from both ends, the state sovereignty system survives to still have the ultimate word over how resources are used. What tools are emerging to influence the Goliath of a state's right to do as it wishes with its environment?

7.5 Dealing with geopolitical conflicts over the environment

James Steinberg, the director of policy planning at the US State Department was quoted in 1995 (*The New York Times* October 9, 1995):

> During the cold war, most security threats stemmed from state-to-state aggression ... now we're focusing more on internal factors that can destabilize governments and lead to civil wars and ethnic strife. We're paying more attention to early warning factors, like famine and the environment.

Former General Secretary of the United Nations, Boutros Boutros-Ghali, wrote in the journal *Foreign Affairs*:

> At the Conference on Environment and Development in Rio de Janeiro in June, 1992, states obligated themselves to take global consequences into consideration in their domestic decisions. This is a fundamentally philosophic undertaking by the world's nations, adding one more pillar to the gradually growing array of internationally accepted principles of national conduct (Boutros-Ghali 1992: 99).

Both statements recognize the need for states to consider environmental issues within the realm of international global security. But how can this need be fulfilled? What mechanisms are available to bring states into compliance?

National-level environmental regulations are becoming more widespread on the planet, albeit with varying degrees of enforcement. But polluters can escape limitations by shifting operations to states with less stringent laws. This factor has been of consideration during negotiations for international agreements such as NAFTA (North American Free Trade Agreement) and in building the EU (European Union). In addition, national laws may build in "grandfather" clauses to avoid penalizing existing polluters, often located in already industrialized states. Should the developing states be subject to more restrictive laws than the wealthier world dealt with historically?

An early attempt to recognize the international nature of environmental degradation, the UN Stockholm Conference, noted in principle 21 that states have the responsibility not to damage the environment within their jurisdiction, but people do have the right to exploit resources within their own borders. Increasing understanding of externality issues has since called into question the extent of this right. In addition, technologies are now allowing for increasing environmental impacts on areas not bounded by sovereign states, such as the upper atmosphere and the deep oceans.

International laws have thus emerged during the twentieth century to constrain the right of a sovereign state to impact the earth biosphere.

BOX 7.8

World heritage sites – threats to sovereignty?

UNESCO, the United Nations Educational, Social, and Cultural Organization, designates places around the world as World Heritage Sites, indicating that they have special value to all humankind and can give land preserves status as Biosphere Reserves. However, in 1996, representatives in the US Congress from Alaska and Montana argued that such a designation for sites in the United States could be a violation of national sovereignty. The United States has 20 of the UNESCO 469 World Heritage Sites and 47 of the World Biosphere Reserves (out of 342 total). Critics of the system complain that it might give the UN pretext for interfering with economic development plans around the sites.

Sometimes these laws take the form of bilateral treaties (e.g. the agreement between the United States and Canada over Great Lakes water quality). Others are multilateral (e.g. the European Convention on Transboundary Air Pollution).

International treaties, or conventions, are usually signed by states at meetings and then require ratification by home legislatures of a sufficient number of states to put the treaty into force. Often conventions set forth principles that then require specific protocols to work out the details, particularly if absolute quantities are involved (such as levels of pollution reduction by target dates).

However, it is still up to states to carry out their promises, and provisions for punishment are a relatively new phenomena. The traditional definition of causing harm to human populations, an action often forbidden under various international laws, now seems to be extended to harming the environment, such as loss of biodiversity or restrictions on pollutants in the upper atmosphere. However, proving liability often requires cooperation of the very states who might then be held accountable. Governments resist admitting liability, as well as sanctions or compensation demands.

7.5.1 THE EARTH SUMMIT

The UNCED (Earth Summit) in Rio de Janeiro resulted in three statements: principles on the environment and development, principles for sustainable management of forests, and principles for general sustainable development. Two international treaties were signed, one on climate change (the greenhouse effect) and one on biological diversity. The United States was initially opposed to the Climate Change Treaty until specific reduction targets were dropped. Some 156 states signed the Convention on Biological Diversity at the conference, but the United States refused and only a later change of political administration in Washington DC resulted in the US signing it. The treaty has been ratified by enough states to put it into force.

The convention acknowledges that maintaining the planet's biodiversity is an international responsibility, although states do have control over their internal resources. It sets a framework for wealthier states to help out the developing world through transfers of technology and money. As noted above, the GEF has a prime responsibility for carrying out this mandate. A major course of conflict during the negotiations had to do with international protection of copyright for any biological resources deriving from the environment. Similar to arguments that emerged during the Law of the Seas Treaty negotiations over mining the deep ocean bed, property rights questions revolved around Western firms having the know-how and investment to develop biological resources, but developing states having ownership claims and wanting royalties. The South has regarded intellectual property law, especially with regard to genetic resources, as a type of immoral transfer of benefits from the earth. The convention that came out of UNCED is unclear about this issue and seems to be trying to satisfy both sides.

7.5.2 INTERNATIONAL TAXES?

In terms of global warming, one alternative to direct mandates for emissions reductions that has been proposed is the use of either carbon taxes or tradable emission permits. This movement is an attempt to commodify, or put prices, on resources previously unpriced, such as clean air. The argument is that the market can be a better distributor of benefits than an international convention. If a

Plate 7.3 Traffic in Pakistan (photo: Kathleen Braden)

global budget for emissions can be established, then budgets of greenhouse-producing gases on a per capita basis could be allocated to states, with a ceiling placed on it to discourage population growth as a way to get more carbon permits. These permits could then be traded on world markets as any commodity. The philosophy is that ownership rights and prices would thus be established, eliminating bad utilization policies and externalities.

Those in favor of this approach argue that it would give developing states the ability to gain the emissions rights they need for economic growth, but provide an incentive for wealthier states to stay with clean technologies. However, problems with these methods are also present: can the norms for emissions be scientifically determined? Who does the monitoring? Would illegal markets develop? Sovereign states would still have the main responsibility for compliance.

7.5.3 ROLE OF INDIVIDUAL CITIZENS

Finally, consumer choices may play an increasing role in protecting the earth ecosystem. New outlooks on consumerism and material well-being in wealthier states may lead to more modest demands on the earth's resources. The rise of "green labeling" to give consumers more information about the environmental history of the products they buy could be helpful. On the other hand, increasingly transnational production of goods, often resulting in products whose origins are difficult to ascertain, may give consumers less control over their choices.

Education and information may be the key. People of fourteenth century Europe could claim ignorance about the cause of the plague that destroyed their security, but modern science continues to reveal valuable information to us about environmental interconnections in our twenty-first century world.

As citizens of the world perceive that their security and well-being are ever more tied in with that of people in other states, even very far removed geographically, actions that might have been seen in the past as self-serving may be revealed to be self-destructive. Conflicts between states over increasingly scarce resources seem moot when impacts of poor management are increasingly global. Human experience has shown that the tragedy of the commons is real: resources owned by everyone and by no one suffer from misuse. But the notion of ownership is an idea that evolved as economic systems developed. Equal evolution and learning is possible as people become more educated about the consequences of our actions. Interdependence is removing all externalities.

8

International Law

KEYWORDS

common morality,
declaratory tradition,
treaty, resolution,
United Nations,
International Court of Justice,
modes of redress,
anticipatory self-defense,
intervention,
mediation,
arbitration,
peacekeeping,
human rights,
refugees,
war crimes,
Nuremberg principles,
UNCED,
UNCLOS

KEY PROPOSITIONS

- International law is inherently based on a notion of common morality, despite cultural diversity at the global scale.
- International law is of increasing relevance to sovereign states, despite flaws and uneven application.
- While less publicized than conflicts, peaceful resolution of geopolitical disputes occurs yearly.
- United Nations intervention in disputes and rights violations has been increasing since the 1970s.
- Individual human rights are increasingly under umbrella protection of international law, despite difficulties in establishing universal values.
- International conventions dealing with use of the earth's resources are impacting the role of sovereign states and private firms and may ultimately affect distribution of economic and political power among states.

8.1 International law and common morality

Law is a system of rules of conduct, established by an authority for the purpose of promoting the values of a society. States are organized to protect their citizens against enemies or actions detrimental to the states' interests; therefore, having some predictability, regularity, and order in their relations with other states is desirable. How can such order be ensured? Can laws be established at the international level that have authority and enforcement power behind them akin to domestic sets of rules? In the absence of an international sovereignty, how can respect for such laws be established?

The moral values of a society are reflected in its laws, since rules arise out of traditions and precedents. Also, law is concerned with creating a sense of obligation and duty, much as moral traditions must do. Scholars of international law sometimes write of the evolution of a common morality among human societies, understanding that such a phenomenon must be highly generalized to allow for the variation of cultures across the earth. In the West, Judeo-Christian traditions that were put forth by scholars such as Thomas Aquinas created an argument for "natural laws", transcending any particular society in time and space. Natural law is derived from reason, logic, and a common sense awareness of what is just and unjust. The Judeo-Christian tradition underlies much of international law, which has developed over several centuries to regulate relationships among sovereign states. Leaders of some non-Western and less developed countries have criticized international law on the grounds that it is culture-bound and imposes Western cultural values on non-Western areas. The moral values behind international law may not pertain, for example, to Marxist states (insofar as much Western tradition in international law derives from capitalism).

Because there is no absolute global authority or police force behind international law, it is sometimes dismissed as a mere symbol. Nevertheless, an examination of the effectiveness of international law reveals the startling truth that for the majority of cases, most states do adhere to international laws. Clearly, states sense some benefit from following a set of moral traditions that contribute to order. Many of the preceding chapters examined the consequences in human lives and environmental destruction when states or other actors determined geopolitical relationships through conflict, instead of through law. In a world that is evermore interdependent and with modern weapons capable of unprecedented destruction, acceptance of the belief that international law can work may be necessary.

8.2 Defining international law

International law has four basic defining characteristics. First, international law obligates a certain moral code of behavior. Because it is constrained by the sovereignty of states and has no international enforcement mechanism on an ongoing basis, international law relies on expectations of behavior. States' actions must show a belief that there will be a mutual benefit in following rules and some consequences in breaking them.

Second, international law is an ongoing process rather than a specific set of rules. It evolves as a body of precedents are created. Scholars have argued that nine fundamental principles of international law have emerged during the twentieth century: sovereignty of equal states, territorial integrity and independence of states, equal rights and self-determination of peoples, non-intervention in internal affairs of other states, peaceful settlement of disputes between states, no use of force or threat to do so, fulfillment in good faith of international obligations, cooperation with other states, and respect for human rights (Nardin and Mapel 1992: 44–45).

Third, international law requires a declaratory tradition. The process of evolving rules must be clearly set forth as treaties, international agreements, resolutions, court rulings, and even custom. In many cases, international laws are a result of conferences that are convened to deal with specific problems (such as global warming or ongoing damage from land mines). Sometimes problems arise when jurists must determine the difference between custom and treaty obligations, especially if treaties are new (such as the UN Law of the Seas Convention) and create new customs, as well as build on old ones. Article 38 of the statute that created the International Court of Justice set forth the sources of international law as: (a) international conventions (agreements); (b) international custom; (c) general principles of law recognized by civilized nations; and (d) judicial decisions and teachings.

Fourth, international law requires a proper authority. Authority can be derived directly from state sovereignty and agreements among states or through international bodies created with the consent of sovereign states (such as the United Nations or European Court of Justice). There may be disagreement, however, as to whether international law is monist or dualist. The monist position regards international law as merely one step up the hierarchy of laws to which a citizen or a society must answer, while the dualist position regards international law as completely separate from sovereign state law. A citizen may choose to argue in national courts that he or she resisted a draft induction or trespassed in protest at a nuclear arms facility to answer to a higher, international law, but such defenses are usually not validated by states' judiciaries.

8.2.1 BRIEF HISTORY OF INTERNATIONAL LAW

International law is a relatively recent phenomenon, developing as the state system emerged out of European political organization of the fourteenth through sixteenth centuries. An increase in trade among governments meant that an expected set of regulations was desirable to keep order. As traditions emerged that sovereignty was vested in rulers over their territory, that states were equal one to another, and that no higher authority could exist over the state, only a body of rules agreed to by consent of the states could be acceptable. In the seventeenth century, Spanish and Dutch jurists developed much of the basis of modern international law in the interests of their states' trading relations. Dutch scholar Hugo Grotius, for example, brought ideas from Judeo-Christian and Roman traditions into laws for states. Much of his writing was used during the twentieth century impetus to create a body of laws governing the oceans, as Grotius had argued that freedom of navigation was an important variable for world trade.

The Swiss scholar, Emerich de Vattel, contributed a classic text, *The Law of Nations*, to the emerging notion of international law in 1758, equating the need for such order with civilization itself in Europe. In fact, the term "international law" did not appear until later centuries, replacing the phrase "law of nations".

By the nineteenth century, Great Britain took the lead in much development of international law. In 1856, the Congress of Paris tried to create a set of rules for maritime warfare, and conferences in Geneva and the Hague discussed conduct of war. By the twentieth century, rules about war proliferated (see below) as did international organizations of presumed authority (e.g. the League of Nations, the United Nations). At the end of World War II, the Nuremberg and Tokyo Trials contributed further thinking about punishment for war crimes and took international law more into the realm of the individual's behavior. Finally, from the 1970s through to the current period, international regulations on many aspects of life, from the environment to human rights to maritime boundaries, have accumulated.

8.2.3 TREATIES AND OTHER STATEMENTS OF INTERNATIONAL LAW

Formal agreements among states are called treaties and these constitute one of the main forms of international law expression. More than 15,000 treaties have been registered at the United Nations. Many are bilateral agreements between two states, such as the 1985 US–Canada Pacific Salmon Treaty. Multilateral treaties (involving more than two countries) have more global impact under international law, and may take the form of conventions (such as the Convention on Biological Diversity that emerged from the UNCED meeting discussed in Chapter 7, or the Geneva Convention on Treatment of Prisoners of War), protocols to set forth more explicit details (for example, the Protocol for Prohibiting in War Asphyxiating, Poisonous, or Other Gases and of Bacteriological Methods of Warfare 1925), charters, or articles of agreement (such as the charter forming the United Nations and the statute forming the International Court of Justice).

The United Nations International Law Commission even created a code for agreements under the Vienna Convention on the Law of Treaties in 1969. Custom dictates that a state first signs a treaty or convention and then ratifies it in domestic legislative bodies. Multilateral agreements then become law (are "in force") when a sufficient number of states have ratified them. After that time, all member states of the United Nations accept them as law, even if they remain not ratified at home (a state accedes to the law). If the ratification process is quite long, states are expected to respect the treaty development process and not undermine the agreements in the meantime. This latter point has been especially contentious in the case of the Law of the Seas, when several industrial states have attempted to formulate alternative treaties governing deep sea mining.

Resolutions are declaratory statements that may also function to create international law. The United Nations General Assembly and Security Council frequently adopt resolutions; examples include the 1985 Security Council Resolution 579 condemning hostage-taking and abduction, or Resolution 661, which in 1990 imposed economic sanctions on Iraq after it invaded Kuwait.

Decisions of the International Court of Justice may also serve as a source of international law, particularly since the law is an accumulation of precedents.

8.3 International law and political institutions

In order for international law to be effective, states must find it in their best interest to comply. Also, there must be the will and the mechanism for states collectively to punish or sanction states that violate international law. In addition to the United Nations, other international bodies (as well as regional associations of states) are creating an impact on accumulating international laws.

8.3.1 INTERNATIONAL LAW AND THE STATE

The enforcement of international law requires states to cede some sovereignty to international organizations charged with enforcement and compliance. Bodies such as the Civil Aviation Organization or the passage of actions such as the 1947 Telecommunications Convention demonstrate that to have a functioning international system, norms must be developed for everything from mail delivery to the sharing of airwaves for transmissions.

International law not only requires the compliance of states, but it also sometimes helps to determine just who is a state. Admission to the United Nations, for example, is often taken as *de facto* evidence of a state's existence, as noted in the argument over the People's Republic of China versus the Republic of China UN seat, the past inclusion of Ukraine and Belorussia (within the USSR) in the General Assembly but not separately in the Security Council, and most recently, the question of Yugoslavia's status.

Slovenia, Croatia and Bosnia-Hercegovina were admitted as new members in 1992. The General Assembly also passed a resolution to suspend the "Socialist Federal Yugoslavia" from participation, but, at the same time, the Serbia-Montenegro federation and provinces (what was left of the old Yugoslavia) was not recognized as the legal successor. Instead, it was invited to apply for new membership. At the time of writing, the issue of whether a "Yugoslavia" still exists was not resolved. The Vienna Convention on the Representation of States in Their Nations with International Organizations of a Universal Character (1975, but awaiting ratification) establishes the privileges and immunities of states' representatives to international organizations.

States' recognition of other states, the establishment of relations, the exchange of ambassadors, and the creation of consulates or embassy missions therefore all have important standing under international customary laws. Diplomats have special status and must be accorded respect, protected against attack, and may not be detained or arrested. Diplomatic immunity means that a states' representatives cannot be pursued under criminal jurisdiction of the state where he or she is posted. The premises of the diplomatic mission are inviolable, as are diplomatic pouches and other communications. Incidents have often sorely tested this custom, however, such as the hostage taking at the United States embassy in Teheran (Iran) in the 1970s, the shooting of a British police officer from the Libyan mission in London, or the case of a diplomat from Georgia who caused a death in Washington DC after driving under the influence of alcohol. The diplomat could be prosecuted under American law only after diplomatic immunity was waived by the government of Georgia.

States also must ensure that violations of international laws are followed by consequences to offending states. Thus, Iraq's 1990 invasion of Kuwait (which was a violation of the United Nations Charter) was first answered by economic sanctions and later by military action. International condemnation in a very public manner at the United Nations, followed by trade sanctions (as in the case of South Africa for its apartheid policies or the USSR for its 1980 invasion of Afghanistan) may serve to bring severe international pressures on a state. However, a case can be made that not all violations are judged similarly or that politics can be kept out of the issues. States, for example, may claim anticipatory self-defense (see below).

The argument for national security may even be cited in cases related to trade sanctions. At the

> **BOX 8.1**
>
> **Preamble to the Charter of the United Nations**
>
> We the peoples of the United Nations determined to save succeeding generations from the scourge of war, which twice in our lifetime has brought untold sorrow to mankind, and to reaffirm faith in fundamental human rights, in the dignity and worth of the human person, in the equal rights of men and women of nations large and small, and to establish conditions under which justice and respect for obligations arising from treaties and other sources of international law can be maintained, and to promote social progress and better standards of life in larger freedom,
>
> And for these ends, to practice tolerance and live together in peace with one another as good neighbors, and to unite our strengths to maintain international peace and security, and to ensure, by the acceptance of principles and the institution of methods, that armed force shall not be used, save in the common interest, and to employ international machinery for the promotion of the economic and social advancement of all people,
>
> Have resolved to combine our efforts to accomplish these aims. Accordingly, our respective governments, through representatives assembled in the city of San Francisco, who have exhibited their full powers found to be in good and due form, have agreed to the present Charter of the United Nations and do hereby establish an international organization to be known as the United Nations.

time of writing, the European Union and other states (such as Canada) have argued that the Helms-Burton Law signed by the US president in 1996 is a violation of international trade policies. The law provides mechanisms for punishing foreign nationals that continue to do business with Cuba, thus bringing the US struggle against the Castro government into the global arena. Countries are pressing a claim against this law before the World Trade Organization, but the US refuses to acknowledge jurisdiction of WTO, citing national security concerns.

8.3.2 INTERNATIONAL LAW AND THE UNITED NATIONS

The United Nations Charter is in itself a multilateral treaty, ratified in 1945. From the original 51 founding members, the United Nations had grown to 185 states by 1997.

The UN was founded on principles such as equality and sovereignty of all members states, peaceful settlement of disputes, and respect for state sovereignty (non-interference). Headquartered in New York City, the UN has grown to encompass an enormous and often financially troubled bureaucracy. Its major organs include the General Assembly (where all members are represented), the Secretariat, the Security Council (with five permanent members – the United States, France, China, Russia, United Kingdom – and ten rotating, non-permanent members), the Economic and Social Council, the Trusteeship Council, and the International Court of Justice (see Table 8.1).

In addition, the UN has programs and other organizations that often govern international regulations, such as the UN High Commissioner for Refugees and the UN Development Program. Specialized agencies in association with the UN also play an important role (e.g. the World Health Organization, the International Monetary Fund, and others; see Table 8.2).

The United Nations is considered a logical successor to the League of Nations. The Security Council may hold the real power in the organization, since only it can create and commit military forces (and a veto by one of the permanent members can end any such movement), but the role of the General Assembly and Secretariat should not be overlooked. The former has often performed an important public function as a world forum to air disputes and a private function for delegates to work out differences behind the scenes. The latter, in the person of a series of strong secretary-generals, has at times intervened to successfully broker cease-fires (such as the case of the Iran–Iraq con-

Table 8.1 Principal United Nations organs and their functions

General Assembly	Composed of all UN members, each with one vote. Controls finances, makes recommendations, oversees and elects some members of other organs. May pass resolutions that contribute to the international law process.
Security Council	Primary responsibility for maintaining international peace and security. May make recommendations and establish peacekeeping forces. Includes 15 members (five permanent – US, Russia, China, France, and United Kingdom – and 10 non-permanent elected for two-year terms). Substantive matters require nine votes. A negative vote ("veto") by any permanent member will defeat a motion. Security Council resolutions are binding on all UN members.
Economic and Social Council	Under the authority of the General Assembly. Coordinates economic and social work of the UN and affiliated institutions. Fifty-four members are elected by the General Assembly for three-year terms.
Trusteeship Council	The five members are the same as the permanent members of the Security Council. Originally oversaw the administration of trust territories. Suspended operations in 1994 with the independence of Palau, the last remaining trust territory. The Council resolved to meet only as the need arose in future.
International Court of Justice	Meets when cases are referred by states and provides advisory opinions to the General Assembly and the Security Council. Made up of 15 members elected by the Security Council and the General Assembly for nine-year terms.
Secretariat	Administers programs and policies of other UN organs. The Secretary-General heads this organ and is elected by the General Assembly based on Security Council recommendations for a five-year term.

The seven secretaries-general have been:

Name	*Country*	*Year started*
Trygve Lie	Norway	1945
Dag Hammarskjöld	Sweden	1953
U Thant	Burma	1962
Kurt Waldheim	Austria	1972
Javier Perez de Cuellar	Peru	1982
Boutros Boutros-Ghali	Egypt	1992
Kofi Annan	Ghana	1997

flict and the work of Secretary Javier Perez de Cuellar, which led to the cessation of hostilities in 1988).

The International Court of Justice (ICJ, often called the "World Court") is made up of 15 members who are elected by a majority of the Security Council and General Assembly for nine-year terms. Usually, member states offer professional

Table 8.2 United Nations organizations

Examples of specialized agencies within the United Nations system and autonomous affiliated organizations

GATT	General Agreement on Trade and Tariffs
IAEA	International Atomic Energy Agency
FAO	Food and Agricultural Organization
IMF	International Monetary Fund
WHO	World Health Organization
IBRD	International Bank for Reconstruction and Development
ICAO	International Civil Aviation Organization
IFAD	International Fund for Agricultural Development
WMO	World Meteorological Organization
WIPO	World Intellectual Property Organization
UNESCO	United Nations Educational, Scientific, and Cultural Organization
ILO	International Labor Organization
UPU	Universal Postal Union

Examples of programs and organs under the auspices of principal organs

General Assembly

UNHCR	Office of UN High Commissioner for Refugees
UNICEF	UN Children's Fund
UNCTAD	UN Conference on Trade and Development
UNEP	UN Environment Program
UNDP	UN Development Program
UNIFEM	UN Development Fund for Women
UNU	UN University

Economic and Social Council

Population Commission, Commission for Social Development, Commissioner for Human Rights, Commission on Transitional Corporations, Economic Commission for Africa, Committee on Crime Prevention, UN Group of Experts on Geographical Names

Security Council

Military Staff Committee, various UN peacekeeping missions (see Table 8.5)

jurists for the positions. The court is only in session when states bring cases, and states cannot be forced into participation. But members of the UN are automatically considered parties to statutes determined by the court. Contentious jurisdiction by the court is a type of binding arbitration, but the court may also make advisory opinions.

By the 1990s, the ICJ was dealing with issues ranging from an allegation of genocide brought by the governments of Bosnia and Hercegovina against Yugoslavia, an advisory opinion about the legality of nuclear weapons (requested by the World Health Organization), an argument between the Pacific state of Nauru and Australia over phosphate profits, to settlement of a fishing grounds dispute between Denmark and Norway, and had decided not to hear cases involving France's nuclear testing in the Pacific (brought by New Zealand) and Indonesia's occupation of East Timor (brought by Portugal). In the last two cases, the ICJ decided it had no jurisdiction. The first woman judge in the 50-year history of the ICJ, Rosalyn Higgins (of the United Kingdom), was seated in 1995.

In addition to the ICJ, other international courts associated with the United Nations are under way. The World Trade Organization includes a Dispute Settlement Body, and the Law of the Sea Convention (in force as of 1994) provides for a new International Tribunal for the Law of the Sea. Separate tribunals were created to deal with war crimes alleged to have been committed in former Yugoslavia and Rwanda.

The ICJ lacks jurisdiction in international criminal cases. As of 1996, a movement was gathering support to create a permanent international criminal court to deal with issues such as terrorism, smuggling, and drug trafficking. The European Parliament and the International Law Commission have been prime movers behind the effort, and in 1998 the statute to create the court was adopted. Such a court is to have 18 judges and its own prosecutor to try cases related to crimes against humanity when national courts decline to do so. Questions have arisen about how much control the UN Security Council should have over the mandate of the court and what would occur if some states refused to ratify a new treaty that created the court.

8.3.3 OTHER INTERNATIONAL ORGANIZATIONS

In addition to United Nations' organs, other international bodies add to the process of international law formation. The European Court of Human Rights (ECHR), formulated as part of the European Union, and the European Court of Justice have both increased caseloads and decisions that contributed to international law. For example, the ECHR ruled against the Spanish domestic courts, which had convicted a member of parliament for an opposition article he had published, and dealt with issues of parents' access to children in foster homes and telephone line tapping for crime prevention.

The Organization of Eastern Caribbean States and the Organization of American States have both been used as organs by the United States to authorize military action (in Grenada and offshore Cuba, respectively) and thus decrease criticism that such interventions were unlawful. The Organization for African Unity has its own mechanisms for conflict resolution among African states (although the record in such conflicts as Chad in the 1980s and Rwanda in the 1990s suggests that its effectiveness has yet to be shown) and has worked to develop African regional law on the environment.

8.4 Conflict resolution

How far should international bodies be willing to go to step in when conflict occurs? Can peacekeeping be accomplished most successfully only when a stronger and superior force enters the fray? What does such intervention mean for state sovereignty?

In the history of UN resolutions to allow intervention into conflicts between states, the euphemism "all necessary means" usually indicates that armed intervention may be sanctioned by the UN and therefore have force of law. However, before that ultimate step is reached, a variety of other tools are available to the global community to defuse violent conflict.

8.4.1 BORDER DISPUTES

Because the integrity of a state's territory is the most obvious and necessary hallmark of sovereignty, violations of borders and claims about conflicting lines on the map may be the predominant cause of interstate conflicts. Amazingly, many more disputes about boundaries are settled peacefully each year than give rise to violence. For example, after years of disputes and even violence in 1995, Ecuador and Peru finally reached an agreement over their border, a process mediated

BOX 8.2

An unusual boundary dispute case: arbitration of the "Pig Wars" between Great Britain and the US

While treaty created the boundary of the United States versus Great Britain along the 49th parallel, the issue of where it would fall at the very western edge (the San Juan Islands) was not to be resolved for many years. Should the international border be along Haro Strait (on the west) or Rosario Strait (on the east)? By the 1850s, the islands contained both American and British settlers, and when a British-owned pig was shot for trespassing in an American-owned garden, it resulted in the locally famous "Pig War" dispute, complete with armed threats and soldiers. Eventually, Emperor William I of Germany was asked to arbitrate. In 1872, he ruled in favor of Haro Strait as the boundary, and the San Juan Islands (now in Washington State) became part of US territory.

Table 8.3 Examples of boundary disputes and settlements

States	*Boundary dispute*	*Mediator*	*Outcome*
Argentina–Chile	Beagle Channel at tip of South America	Vatican 1984	Chile gets islands of Lennox, Picton, and Nueva, but Argentina gets increased maritime rights
Chad–Libya	Aozou Strip; Libya claimed never demarcated properly and therefore it had rights to territory	International Court of Justice, 1994	Ruled that border was set by 1955 treaty between Libya and France, and Libya has no claim
El Salvador–Honduras (border war had occurred earlier in 1969; states submitted to arbitration in 1986 to prevent further war)	Six land parcels along the Honduras–El Salvador border and three islands in the Gulf of Fonesca as well as the Gulf itself	International Court of Justice, 1992	Honduras received two-thirds of the territory; El Salvador one-third. Gulf to be shared between the two states and with Nicaragua; two islands (Meanguera and Meanguerita) to El Salvador; El Tigre Island to Honduras

by the United States, Argentina, Brazil, and Chile (for other examples, see Table 8.3 and Figure 8.1).

Evan Luard, a research fellow at Oxford and member of the British foreign service in the 1950s created a taxonomy of border disputes (Luard 1970: 14–15):

1 No recognized boundary or treaty exists; therefore, conflict is over traditional and maybe arbitrary claims (e.g. European powers' claims in Africa; Saudi Arabia versus Abu Dhabi and Muscat over boundaries on the Arabian peninsula; international seabed claims).

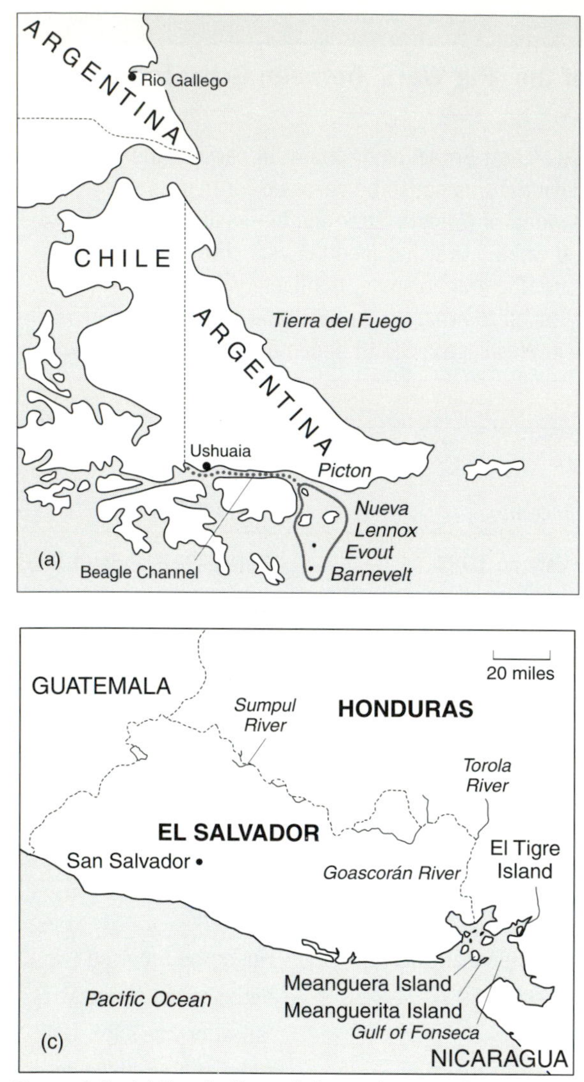

Figure 8.1 (a) Beagle Channel dispute between Argentina and Chile. (b) Aozou Strip dispute between Chad and Libya. (c) Honduras–El Salvador border disputes

2. A *de facto* frontier exists, but legitimacy is challenged by one side (e.g. Germany versus Poland before World War II; Somalia versus Ethiopia over Ogaden).
3. Two rival delimitations exist (e.g. China versus India in the Himalayan Range).
4. There is a mutually agreed delimitation of a frontier, but a dispute about demarcation on the ground (e.g. Cambodia versus Thailand over Preah Vihear Temple; Chile versus Argentina over Beagle Channel).

Types of boundaries may also contribute to international conflict. As noted in Chapter 5, superimposed boundaries create a situation ripe for ethnic conflict as communities of interest are divided or artificially united with other communities. Cultural and political changes may shift around boundaries, decreasing their stability, and the discovery of exploitable resources in a region may increase tensions over boundaries (such as the China Sea dispute discussed in Chapter 7).

One example of a continuing dispute over boundaries and cultural borders is the international boundary between Greece and Turkey. From 1453 to 1832, an independent Greek state did not exist, but the Ottoman Empire dominated the region. In 1832 an independent state was established, but the situation was complicated by the fact that some Greeks had converted to Islam and others remained Orthodox.

Great Britain gave control of the Ionian Islands to the Greek state; in 1913 Macedonia and other Aegean islands, Crete and in 1923 western Thrace came under Greek control. The 1923 Treaty of Lausanne created a permanent border, but Turkey objected to the proximity of the international border to its coastline (see Figure 8.2). This issue

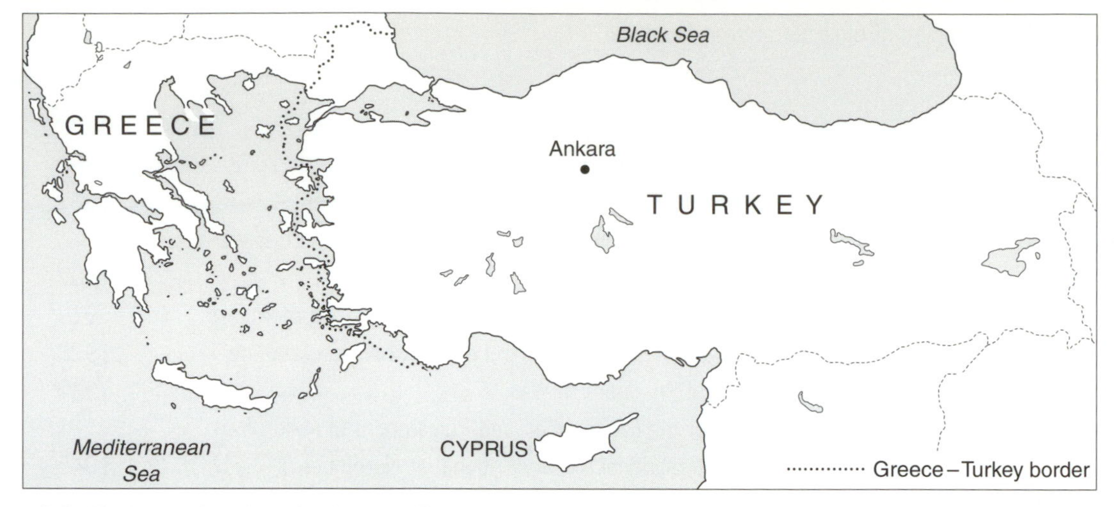

Figure 8.2 The international boundary between Turkey and Greece extends almost to the shore of Turkey

remains a raw wound between the two states today, compounded by the split Turkish and Greek heritage population on Cyprus.

8.4.2 MODES OF REDRESS SHORT OF WAR

Intermediary steps between outright violent conflict and peaceful settlement of disputes also exist in international relations. These may be referred to as modes of redress when one state has a complaint against another.

Retorsion is an unfriendly but pacific act against another state in response to a perceived wrong over borders or other disputes. It may include a state enacting special tariffs, immigration restrictions, or the withholding of aid. The United States, for example, each year certifies states as cooperative or not in the battle against illegal narcotics traffic. States viewed as non-cooperative may lose foreign aid.

Retaliation is a somewhat more serious signal and may involve a trade embargo (such as the UN action against South Africa over apartheid) or a boycott. This last tool has been used particularly over the Olympic Games as a symbol of complaint. In 1956, Egypt, Lebanon and Iraq boycotted the games in Melbourne, Australia, in protest over the United Kingdom–France seizure of the Suez Canal. In 1980, the United States and 40 other states boycotted the games in Moscow to protest against the USSR invasion of Afghanistan. Retaliation can become even more heated when ambassadors are recalled, embassies closed, and diplomatic relations broken off.

Reprisals are short-term aggressive acts of protest. These include brief trans-border incursions and seizures of ships. The legality of such measures has been a matter of debate under international law, as noted below in the Mayaguez incident involving the United States and Cambodia or the United States bombing of Libya in 1986 in reprisal for alleged terrorism support.

8.4.3 LAWS OF WAR

Is war legal under international law? This question has been a matter of heated argument over many centuries as humanity has made repeated attempts to outlaw war. Yet, the issue is complicated by definitions of exactly what constitutes just and unjust war, aggression versus self-defense, or even lawful conduct during war.

The peace movements in the nineteenth and early twentieth centuries and the horror of World War I contributed to a growing belief that war should be illegal as well as immoral under international norms. In 1919 at the Paris Peace Conference, the League of Nations was established with the aim of restricting use of force by its covenant member states. Article 12 required arbitration of disputes and Article 15 created an obligation not to go to war. But states could turn

Table 8.4 Examples of multilateral treaties on war

Agreement	*Year signed or drafted*
Geneva Convention on Treatment of Wounded Soldiers	1864
St Petersburg Declaration on Use of Projectiles of Certain Size During War	1907
Hague Convention Respecting Rights and Duties of Neutral Powers and Persons in Case of War on Land	1907
League of Nations Article 16 – sanctions of pacific settlement of disputes among members	1919
Treaty for the Renunciation of War (Kellogg–Briand Pact) within League of Nations regime	1928
Geneva Convention Relative to Treatment of Prisoners of War	1949
Agreement of UK, US, France and USSR for the Prosecution and Punishment of Major War Criminals of the European Axis (created International Military Tribunal at Nuremberg)	1945
Charter of the United Nations, Article 2	1945
Treaty on the Non-Proliferation of Nuclear Weapons	1968; renewed 1995
United Nations General Assembly Definition of Aggression Resolution	1974
United Nations Security Council Resolution on Somalia (allowed US troops to be sent)	1992
Convention on the Prohibition, Development, Production, and Stockpiling of Chemical, Bacteriological, and Toxic Weapons	1993
Convention on the Prohibition of the Use, Stockpiling, Production, and Transfer of Anti-personnel Mines and on Their Destruction	1997

to war if they felt there was no other recourse after a period of three months moratorium on fighting. The Treaty Providing for the Renunciation of War as an Instrument of National Policy (known as the Kellogg-Briand Pact) was signed in 1928 to further restrict violence (see Table 8.4). War was only permitted in the case of self-defense.

The creation of the United Nations offered another opportunity to mitigate the recourse to war. Article 2, Paragraph 4 in Chapter I of the Charter states: "All members shall refrain in their international relations from the threat or use of force against the territorial integrity or political independence of any state, or in any other manner inconsistent with the Purposes of the United Nations." But Article 51 goes on to state that force is allowable under certain circumstances. The exceptions are: (1) force in self defense, (2) force by authorization of UN Security Council, (3) force undertaken by five major powers before Security Council was functional, and (4) force undertaken against enemy states of World War II.

Chapter VII, Article 39, allows the Security Council to determine what constitutes a threat to peace and to take measures to restore international peace and security. As noted below, this power has been increasingly invoked by the Security Council in cases of intervention. Article 39 also empowers the Security Council to order states of the UN to use force against transgressor states should the need arise. It also provides a mechanism for the Security Council to impose sanctions. Since the UN has no standing army, Article 43 says states must make available contingents of their armed forces that the Security Council could summon. Further, a Military Staff Committee can be formed under Article 47 and is composed of the Chiefs of Staff (or their representatives) of the Security Council's permanent members.

An early example of the use of this article

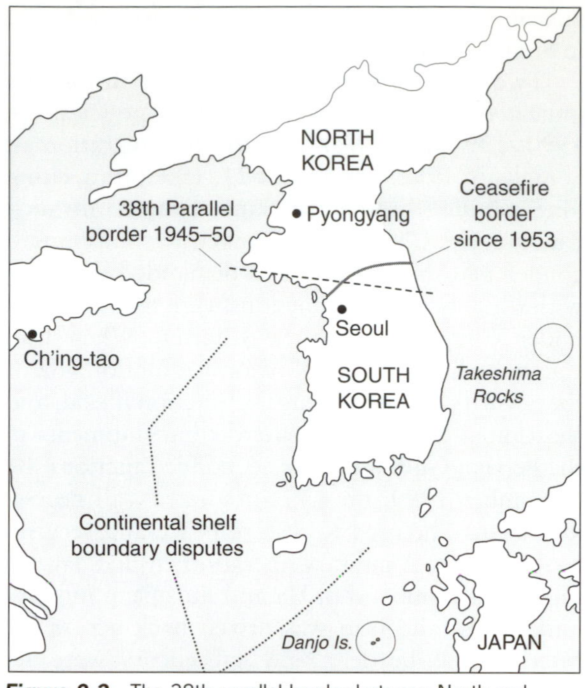

Figure 8.3 The 38th parallel border between North and South Korea and the 1953 cease-fire line

occurred in 1950 when North Korean forces invaded South Korea (Figure 8.3). The United States brought a complaint to the Security Council. Because the USSR had been boycotting the Council to protest the seating of Nationalist China, there was no permanent member in opposition to exercise a veto. The Council adopted Resolution 82, which called on member states to assist the UN with commitment of troops, Resolution 83, which determined that North Korea had indeed violated international peace and authorizing use of force against it, and Resolution 84, which set up a Unified Military Command. The conflict lasted from 1950 until 1953 when an agreement established a demarcation line and mechanism for exchanging prisoners of war.

One question that frequently arises is the meaning of self-defense, since the UN Charter allows for use of force in such cases. Of particular complexity is what constitutes anticipatory self-defense. Must a state wait to be attacked first before it can exercise self-defense and be within its rights under the UN Charter?

Several famous cases involving Israel have arisen over the years. In 1967, Israel attacked forces of the United Arab Republic, believing that all evidence pointed to an imminent attack on Israel due to gathering of Arab forces on the Sinai Peninsula. Subsequent debate ensued about which side actually initiated the June war of that year. In 1981, Israel (claiming its right to self-defense and anticipating Iraqi ability to construct nuclear weapons) sent an air attack into Iraqi territory to destroy the Osarik nuclear reactor being constructed near Baghdad. Israel also conducted a raid into Tunisian territory against the relocated Palestinian Liberation Organization headquarters in retaliation for attacks on Israeli citizens.

In 1962, the United States argued self-defense in its blockade of Cuba to interdict further shipment of Soviet short-range missiles into that state. The USSR argued collective security when it formulated the "Brezhnev Doctrine" as its right to intervene when allied socialist states, such as Czechoslovakia or Afghanistan, were threatened.

The laws governing proper conduct of war were enlarged in scope when the issue of war crimes became prominent after World War II. The establishment of military tribunals to try leaders in Germany and Japan created precedents which are often cited. Norms of international law continue to develop to determine what actions are permissible in the conduct of a war and what should be considered as criminal behavior. Such laws have been quite controversial because the victorious side usually sits in judgment on the losing side and determination of illegal war conduct is thus skewed. For example, some argued that aerial bombing of non-military targets by the Allies during World War II also constituted crimes against humanity (see Section 8.8).

8.5 "All necessary means": intervention in conflict

Multinational organizations and single sovereign states may intervene in conflicts between or even within states, but the pretext for doing so under international law is complex. The most benign form may be the offer of pacific settlement of disputes, especially if other modes of redress are exhausted and violence seems imminent. States may choose bilateral methods of negotiations without outside intervention to assist, but once third parties are invited in, three forms of assistance may be sought:

1 good offices, in which the third party merely provides a location and support services for the disputing states;
2 mediation, in which the third party offers solutions but acts more as go-between (e.g. the United States' role in mediating the dispute between Israel, its Arab neighbor states, and the Palestinians);
3 arbitration, especially binding arbitration, in which the parties agree ahead of time to let an outside party solve the dispute; states consenting to use the ICJ to arbitrate disputes would be an example.

Traditional international custom prior to creation of the United Nations held that states could aid other states when insurgencies or low-level conflict occurred, but once the violence escalates to actual war, the interfering state would no longer be considered neutral.

8.5.1 PEACEKEEPING BY THE UNITED NATIONS

The advent of the United Nations meant that intervention has become increasingly codified. From the 1950s until the 1970s, the United Nations mainly served to monitor settlement of conflict, observe frontiers, and enforce cease-fires (see Table 8.5). Peacekeeping missions are authorized by the United Nations, but during the Cold War, most proposals to establish peacekeeping forces were vetoed by one or more members of the Security Council.

The original idea of UN peacekeeping operations was that use of force should only be in self-defense, the consent of the parties (states) involved in the dispute was required, and that peacekeepers must maintain impartiality.

With the end of the Cold War in the late 1980s, the number of UN peacekeeping missions increased. By 1994, 70,000 peacekeeping personnel, including troops, police, and observers, were under UN auspices. The UN now performs intervention functions as varied as monitoring elections (e.g. in Namibia, Angola and Cambodia), creating safe havens (e.g. in Croatia and Bosnia), and guarding weapons that are decommissioned (e.g. in Somalia and former Yugoslavia).

Two events in particular during the 1990s indicated a shift in UN intervention policies: the 1990–1991 Gulf War and the disintegration of Somalia in 1992. On August 21, 1990, Iraq, citing long-standing irredentist land claims, invaded Kuwait. The UN Security Council Resolution 660 condemned the invasion and demanded that Iraq withdraw. Economic sanctions were then imposed under Chapter VII of the Charter. When Iraq claimed it had annexed Kuwait, Resolution 662 of the Security Council deemed the action illegal, and the United States began interdicting shipments in the Persian Gulf under the economic sanctions. By November 1990, the UN authorized "all necessary means" be used to restore peace and security to the region. A military operation ensued, largely under the command of US and European military authorities, and Iraq was driven back across the international border. New precedents were set under subsequent resolutions (April 1991, Resolution 687), which set the cease-fire conditions and compelled Iraq to destroy weapons stores, under light of international inspection. Resolution 688 in April of 1991 dealt with the internal repression of citizens in the country. Iraq subsequently gave up claims to Kuwait, but a no-fly zone was established for Iraqi aircraft over parts of its territory and economic sanctions continued until 1996. At the time of writing, sanctions against Iraq selling its oil resources were beginning to be lifted, but the US had mitiated a bombing campaign to force admission of weapons inspectors.

The UN action in Somalia was equally precedent-setting for intervention norms. In 1991, the president of Somalia was deposed and civil disintegration followed as clans engaged in warfare. The UN at first sought to provide humanitarian assistance to the suffering Somali populations, but protection of relief supplies was problematic. In December of 1992, the Security Council voted in Resolution 794 to authorize "all necessary means" (i.e. a military intervention) to ensure relief delivery. Troops from the United States were initially sent in, later officially as part of UNOSOM peacekeeping force. However, by 1993, fighting continued and soldiers of UNOSOM were killed (including the killing of US

Table 8.5 Examples of United Nations peacekeeping and observer missions (if mission concluded, ending date shown as of 1996)

Mission	*Abbreviation*
UN Truce Supervision Organization 1948	UNTSO
UN Military Observer Group in India and Pakistan 1949	UNMOGIP
UN Peacekeeping Force in Cyprus 1964	UNFICYP
UN Disengagement Observer Force 1974 (Golan Heights)	UNDOF
UN Interim Force in Lebanon 1978	UNIFIL
UN Iraq–Kuwait Observation Mission 1991	UNIKOM
UN Observer Mission in El Salvador 1991	ONUSAL
UN Mission for the Referendum in Western Sahara 1991	MINURSO
UN Protection Force 1992	UNPROFOR
UN Transitional Authority in Cambodia 1992	UNTAC
UN Operation in Somalia 1992–95	UNOSOM
First UN Emergency Force 1956–67 (Suez crisis)	UNEF 1
UN Observer Group in Lebanon June–December 1958	UNOGIL
UN Operation in the Congo 1960–64	ONUC
UN Security Force in West New Guinea 1962–63	UNSF
UN Yemen Observer Mission 1963–64	UNYOM
Sec. General Rep. in Dominican Republic 1965–66	DOMREP
UN India–Pakistan Observation Mission 1965–66	UNIPOM
Second UN Emergency Force 1973–79 (Israel–Egypt)	UNEF II
UN Good Offices Mission in Afghanistan and Pakistan 1988–90	UNGOMAP
UN Iran–Iraq Military Observer Group 1988–91	UNIIMOG
UN Angola Verification Mission 1989–91	UNAVEM
UN Transition Assistance Group 1989–90	UNTAG
UN Observer Group in Central America 1989–92	ONUCA
UN Angola Verification Mission II 1991	UNAVEM II

soldiers quite publicly before the world's news media). In 1994, the United States decided to withdraw from the operation, and, at the time of writing, the UN has still not been able to achieve its objectives in Somalia.

The issues that created a precedent for intervention in Somalia were the same ones that may have caused the mission to fail. At the time of the intervention, Somalia had no functioning government, and therefore the UN principle that intervention should include the consent of the state could not be met. The situation there was later echoed in Rwanda in the mid-1990s, when internal disintegration of order and outright genocide made determination of the responsible state government difficult.

Such interventions by the United Nations will be more and more difficult to accomplish successfully if warfare is increasingly caused by ethnic strife and state devolution. It may be that the mechanisms of the UN to ensure peace and stability among states cannot be extended to intervention into awkward internal wars. Yet, the growing presence of the world media eye and public awareness of human suffering associated with such wars places continued pressure on governments to "do something". Intervention is therefore often claimed on humanitarian grounds, and yet there are few occasions in which purely apolitical, humanitarian intervention really occurs. If international law is indeed a process, rather than set rules, then dealing with norms for such tragedies may be its greatest challenge in the next century.

8.5.2 UNILATERAL INTERVENTIONS BY STATES

States often claim military intervention in another state's territory based on humanitarian grounds. One of the prime pretexts for such incursions is the rescue of nationals. Since UN Charter Article 51 allows for self-defense, states have argued that this may extend to their protection of citizens outside national boundaries.

Once again, Israel provides the most applicable case study of this phenomenon. In 1976, Palestinians hijacked an airplane en route to Paris and forced a landing in Uganda. Israel sent a commando unit (the now-famous "raid on Entebbe", after the airport title), which killed the Palestinians and freed the Israelis held hostage. Uganda did not give permission, and some Ugandan soldiers were killed in the action. Israel claimed self-defense to the Security Council and that the Ugandan government had been preventing progress.

Many cases of this type have occurred since the UN was formed; the earliest being in 1946 when Great Britain threatened to send in troops to help British nationals during riots in Iran. A famous case involved the US merchant ship, *Mayaguez*, which was taken by Cambodia in 1975 (who cited that the ship had been in its territorial waters) near an island that had been claimed by both Vietnam and Cambodia. Using self-defense as an argument, the United States conducted air strikes and finally freed the crew through a land-based military action. The action was not without its critics at the United Nations, who claimed that negotiations to release the men had been under way and the US too impatient.

Counter-intervention is another justification used by states to intervene in disputes outside their boundaries. The United States, for example, argued counter-intervention for its policy in Central America. Claiming that Nicaragua had intervened to help insurgents in El Salvador, the US placed mines in Nicaragua's harbors. The case went before the ICJ, but the US refused to recognize the court's decision.

The United Nations is both expanding its activities and dealing with internal management and financial difficulties simultaneously. The relationship between the United States and the UN has also evolved through many stages. Initially, the US was a key founding state for the institution. In the 1980s, the US entered a period of skepticism about the UN, often resulting in failure to provide financial support and it pressured members not to re-elect Boutros Boutros-Ghali as Secretary-General in 1996. But the US has also had its way more frequently on the Security Council with the weakening of Russia's role. New power relationships have been forming at the UN as developing states prove increasingly influential and as China emerges as a world-level power.

8.6 Human rights

Many of the activities of United Nations agencies are underlain by concerns for human rights. Judge Rosalyn Higgins of the International Court of Justice wrote in her book, *Problems and Process: International Law and How We Use It*, that "human rights are rights held simply by virtue of being a human person" (Higgins 1994: 96). She notes that they are inherent rights that cannot be given or withdrawn at will by any domestic legal system. While scholars may argue over whether there can ever be a universal set of standards (a common morality), the oppressed of the world's states may wish that the benefits of outside assistance might come more quickly.

Since much international law is designed to govern relationships among states, the idea that it can spell out obligations owed directly to each individual person may appear unusual. Yet, international law now holds individuals accountable (see Section 8.8) and, increasingly, individual persons can gain access to international tribunals. Several complicating factors arise with the growing trend toward international human rights laws:

1 How should variations in cultural values be accounted? For example, universal rights are

now explicit for children and for women, yet, in the case of women, cultural traditions associated with such practices as female circumcision create complexities in applying universal standards. Laws regarding adoption, inheritance, and other personal considerations face similar problems.

2 How can the rights of individuals be equated with states' rights, especially in the case of ethnic minorities and self-determination if it erodes a state's sovereign territory? The United Nations is committed to both the territorial integrity of its members and rights of self-determination for all peoples.

3 Is the emphasis on individual rights an artifact of the western world? For example, how should rights to own property be accounted for in socialist or communist states?

Yet while such points are debated, states have committed horrendous acts against their own people within state internal borders, indicating that some common principles of rights are demanded, such as the right to security from state-sanctioned violence. The treatment of indigenous peoples under colonial rule, Native American groups within the United States, Cambodians under the Pol Pot regime, Soviet citizens and ethnic minorities under Stalin, Tutsi and Hutu Rwandans under various regimes, East Timorese by the government of Indonesia, and ultimately, European Jews by the National Socialist government of Germany, give some indication of all the number of deaths that may occur without classic interstate conflict.

In 1948, the United Nations General Assembly adopted the Universal Declaration of Human Rights, with 30 articles that set forth standards of rights for every human being (see Table 8.6). Additionally, various human rights covenants have also been adopted, such as the International Covenant on Civil and Political Rights and the International Covenant on Economic, Social, and Cultural Rights (both in 1966). International covenants or declarations have also been accepted on the elimination of racial and sexual discrimination. Within the United Nations there is now a High Commissioner on Human Rights, as well as one on Refugees; and, in 1993, the UN sponsored an international conference on human rights, held in Vienna.

Along with various UN organs and the European Court of Human Rights noted above, other international bodies have been concerned with abuses of human rights. The Conference on Security and Cooperation in Europe (CSCE) issued its final act in 1975 and devoted Section VII to "Respect for Human Rights and Fundamental Freedoms, Including the Freedom of Thought, Conscience, Religion or Belief" and Section VIII to "Equal Rights and Self-Determination of Peoples". Subsequent to this document (also called the "Helsinki Accords"), the CSCE has continued to monitor and issue reports on human rights in European states.

Of particular significance to state sovereignty has been the issue of minority rights. The CSCE published a report in 1991 calling for the protection of minority rights but reaffirming the territorial integrity of states, and subsequently established a High Commissioner on National Minorities in 1992. The Council of Europe has been attentive to issues of linguistic minorities within European states and issued declarations to support the preservation of language diversity. In 1992, the UN General Assembly adopted the Declaration on the Rights of Persons belonging to National or Ethnic, Religious and Linguistic Minorities.

UN statements on human rights do not necessarily argue that all minority groups within states have a right to secede; in fact, the UN Charter adheres to the principle of state sovereignty over territory. The initial thrust of the UN after its creation was to push for decolonization. The human right of self-government is favored by UN declarations, including the rights of minorities within states to fair representation, government participation, and non-discrimination. Yet international law does not argue against adjustment of boundaries or even the formation of new states, if done in a lawful manner (such as the split of Czechoslovakia into two new states).

Refugee populations create special issues within human rights laws. In 1974 there were an estimated 2.4 million refugees in the world. By 1995 the number was estimated at 23 million. If internally displaced people are included, the figure

Table 8.6 Human rights as set forth by Articles 3 through 28 of the United Nations Universal Declaration of Human Rights (General Assembly Resolution 217, adopted December 10, 1948)

- Right to life, liberty, and security of person
- No one shall be held in slavery
- No one shall be subject to torture, cruel, inhuman, or degrading treatment or punishment
- Right to recognition everywhere as a person before the law
- Right to equal protection before the law without discrimination
- Right to remedy by national tribunals for acts violating fundamental rights
- No arbitrary arrest, detention, or exile
- Right to fair and public hearing of any criminal charge
- Right to be presumed innocent until proven guilty in public trial with guarantees of defense
- No one to be held guilty of an offence that did not constitute a penal offence at the time it was committed
- No arbitrary interference with privacy, family, home, correspondence, nor attacks upon honor or reputation
- Right to freedom of movement and residence
- Right to leave and return to own country
- Right to seek asylum in other countries
- Right to a nationality
- Men and women of full age have the right to marry and found a family; marriage only with consent of spouses; family due protection by the state
- Right to own property alone and in association with others
- No one shall be arbitrarily deprived of his property
- Right to freedom of thought, conscience, and religion
- Right to freedom of expression
- Right to peaceful assembly and association
- Right to take part in government
- Right of equal access to public service
- Right to genuine elections and equal suffrage
- Right to social security
- Right to work, to free choice of employment, to just working conditions, to protection against unemployment
- Right to equal pay for equal work
- Right to form and join trade unions
- Right to rest and leisure
- Right to adequate standard of living
- Right to education, including free elementary education
- Right to protection of moral and material interests resulting from scientific, literary, or artistic production of which person is author
- Right to a social and economic order in which these rights and freedoms can be realized

may reach almost 50 million people, or one of every 114 people in the world, according to the UN High Commissioner for Refugees. The human right to political asylum is sorely tested as refugee numbers skyrocket, particularly in Africa. Ethnic conflict, state wars, and, increasingly, environmen-

Plate 8.1 Palestinian refugee camp, Gaza (photo: Kathleen Braden)

tal crises have created whole new populations of displaced people, demanding protection and basic human services. In many cases, the inheritance of superimposed boundaries continues to propel often violent adjustment of populations.

8.7 Keeping order beyond interstate conflict

As noted in Chapter 6, conflict in the modern world is not limited to classic wars between states. Ethnic strife, terrorism, environmental disruption, and international crime are all destabilizing phenomena that also require responses from international governing bodies.

8.7.1 INTERNATIONAL LAW AND TERRORISM

The mere definition of such problems as "terrorism" make the completion of international laws difficult. The United States Court of Appeals in 1984 ruled against considering an attack on an Israeli bus in 1978 as a "violation of the law of nations". The court cited the fact that little world consensus existed as to what constitutes terrorism, and therefore it would be difficult to determine which international laws had been broken.

In 1970, one of the first multilateral agreements against terrorism was adopted at the

Plate 8.2 United Nations Relief and Words Agency (UNRWA) for Palestine refugees in the Near East (photo: Kathleen Braden)

Hague: the Convention on the Suppression of Unlawful Seizure of Aircraft (Hijacking). It calls upon states where the perpetrators are found to prosecute or extradite. Additional conventions were adopted by the UN General Assembly to enhance the protection of diplomats (1973), against the taking of hostages (1979), and in 1985, UN General Assembly Resolution 61 was passed to "prevent international terrorism" (although this was a largely symbolic action with little definition of what exactly constitutes terrorism).

However, terrorist incidents must either involve violations of domestic laws (e.g. against homicide) or, if they are to be considered a type of warfare, then international laws calling for the protection of non-combatants are most often violated by terrorists, who largely target civilians.

8.7.2 INTERNATIONAL LAW AND GLOBAL ORGANIZED CRIME

Despite the growing global nature of crime rings, most criminal activities are still covered by national laws. However, worldwide crime can spill over into the international law arena in several ways:

1 Illicit banking and industrial activities may occur in violation of UN trade sanctions. Evidence suggests that Russian and Serbian companies operating on Cyprus, for example,

may have been shipping supplies into former Yugoslavia against international laws.

2 Organized crime smuggling of plutonium violates rules of the International Atomic Energy Agency and the Nuclear Non-Proliferation Treaty. Both the German government and the United States Energy Department have maintained in recent years that Russia's stockpile of nuclear materials is increasingly compromised.
3 Crime associated with international financial circles and information transfer can be in violation of World Trade Organization (and other international economic institution) rules.

The United Nations (which already had a Committee on Crime Prevention and Control within the Economic and Social Council) recognized the threat to international peace and stability from organized crime, and in 1994 held a conference of 138 states, meeting in Naples. The conference was a first step in creating new international conventions on prosecution of organized crime perpetrators. A similar convention already exists in the cases of piracy and illegal narcotics trafficking. The UN Drug Control Program operates out of Vienna to coordinate international efforts to stop illegal narcotics. In addition to the UN efforts, the European Union Europol anti-crime agency began operations in 1994, but the agency faces uneven national laws within Europe.

8.8 International law and the individual

Problems in determining international standards for human rights and criminal activities demonstrate that international law still treads lightly in areas of state sovereignty. However, under international law participants may range from states to non-governmental organizations, businesses, and even to individuals. Traditionally, individuals had to bring claims against states via their own state government, rather than directly, but the advent of human rights laws and the actions of the Nazi government during World War II changed this view of the individual. Both responsibility for actions and rights as a human being were reinforced after 1945.

At the end of World War I, individuals were usually not liable for violations of international law unless national laws were also broken. However, some exceptions had already occurred in history. In 1479 Peter of Hagenbach, a governor in Breisach (now in Germany), was tried by an international set of judges from Austria, Switzerland and various cities for criminal treatment of his population. Despite his defense that he was only taking orders, he was sentenced to death. In 1815 the Congress of Vienna charged Napoleon with violating international agreements

BOX 8.3

Some strange ways in which international law affects individuals

In the 1970s, a wealthy American widow was murdered on a cruise ship sailing in the Strait of Juan de Fuca between the United States and Canada. Who had jurisdiction over the suspect? Originally, the US claimed it, because the ship was on the US side; then it was ruled to have been on the "high seas"; finally, it was placed in Canadian waters, giving the Canadian government the right to try her traveling companion for bludgeoning her to death with a bottle of champagne.

In 1988, Alfred Nasseri traveled to France from Iran by way of Belgium. He had been born in Iran when it was under British jurisdiction and had been a student in Great Britain. When he returned to post-Shah Iran, he was arrested and his passport taken away. He could leave the country, but not return. Nasseri went to Belgium and was given asylum by the UN High Commissioner for Refugees, but his papers, which would have allowed him to choose a European citizenship, were stolen with his suitcase in Paris. He attempted to travel to Britain, but without a passport, was returned to the airport in Paris. As of 1995, he was still there, a "citizen" of de Gaulle airport, since France would not admit him without a passport. On May 20, 1995 the *Los Angeles Times* reported, "Nasseri – who has adopted the nickname 'Alfred' – has become a fixture in the circular confines of Terminal One, well known to shopkeepers, flight attendants, police officers, and even some passengers."

and ultimately banished him. After the US Civil War, the commandant of the infamous Andersonville Prison was charged with crimes for his treatment of Union prisoners.

Prosecution of individuals for crimes against international laws regarding war are usually kept to leaders or people in authority, rather than individual soldiers. National military courts, however, may use international laws as well as domestic for prosecuting war crimes committed by soldiers. (For example, the case in which US soldiers serving during the Vietnam war in 1968 massacred 500 civilians in the village of My Lai. Lieutenant William Calley was subsequently convicted under court martial, sent to prison, and later released when the conviction was overturned.) The argument is that leaders are best able to mitigate against unlawful actions of states and therefore are particularly responsible for the illegal conduct of states.

The Nuremberg Principles of holding leaders accountable for war crimes began with the creation of the 1945 International Military Tribunal by France, the United Kingdom, Russia and the United States. The document that created the tribunal was key to subsequent interpretations of crimes. It defined transgressions as:

(a) Crimes against peace: planning and waging a war of aggression in violation of international treaties.
(b) War crimes: violations of war or customs of war. Murder, ill treatment or deportation to slave labor of civilians, ill treatment of prisoners of war, killing hostages, plunder of public property, wanton destruction of places not justified by military necessity.
(c) Crimes against humanity: murder, extermination, deportation against civilian populations, persecution based on political, racial, or religious affiliation.

In the opening statement of the tribunal, Judge Robert Jackson stated, "The wrongs which we seek to condemn and punish have been so calculated, so malignant and so devastating, that civilization cannot tolerate their being ignored, because it cannot survive their being repeated." Twenty-one Germans were tried at the Palace of Justice in Nuremberg, Germany. Evidence was provided to show that Nazi Germany had long planned aggressive warfare against its neighbors, and graphic information about concentration camps was provided. Three defendants were acquitted, eight sentenced to prison terms, and ten sentenced to hang.

Plate 8.3 Monument to victims of Stalin terrors, Siberia (photo: Kathleen Braden)

However, questions remained after the Nuremberg and Tokyo War Crimes trials at the end of Word War II. Subsequent release of war documents showed that the Allied leaders deliberately chose to target German cities and civilian populations for aerial bombing campaigns to create psychological impact. Was this also not a violation of international laws protecting civilians? United States firebombing of Japanese cities and the use of nuclear weapons on Hiroshima and Nagasaki was also of questionable moral and legal validity, despite arguments that such actions hastened the end of the war.

Still, a better focus on the crime of genocide did arise out of the World War II tribunals. In 1948 the UN. adopted Resolution 260, the Convention on the Prevention and Punishment of the Crime of Genocide. The phenomenon was defined as acts "committed with the intent to destroy a national, ethnic, racial, or religious group." The convention went into effect in 1951, although it was not ratified by the United States until 1986.

War crimes tribunals have recently been established in Rwanda and for the former Yugoslavia to bring individuals to justice. The efficacy and apolitical nature of both tribunals, however, have been questioned. The establishment of a permanent international criminal court may eventually nullify the need to create separate tribunals.

8.9 UNCLOS and UNCED: the wave of the future?

International law deals not only with transgressions committed by states and individuals, but also with the creation of an orderly world system. Thus, much international law, regulation, and norms deal with everyday matters of commerce, communications, and international business. The issue of resource use is fast becoming an important direction in international law and may affect the balance of power on the planet.

Two examples of such international laws are the conventions that resulted from the United Nations Conference on Law of the Seas (UNCLOS) and the United Nations Conference on the Environment and Development (UNCED; see Chapter 7).

Planning for the creation of a set of international laws to govern use of the world's oceans began in the 1950s at the United Nations. The initial impetus came from the United States and other large maritime powers who desired a stable system of boundaries for commerce and military use of the oceans. Some progress was made, but it took two more massive UNCLOS sessions before a document was adopted in 1982 by the conference. Issues such as the width of the territorial sea, the creation of an exclusive economic zone (see Figure 8.4), rights of coastal, landlocked, and archipelagic states, pollution of the seas and scientific research, and regulation of trade were basically resolved by the end of the conference. However, the issue that was to prove most problematic was the exploitation of resources from the deep seabed.

Ultimately, the United States did not sign the final act, but by 1994 enough states had signed and ratified the convention that it entered into force of international law. The United States and other industrial countries with multinational mining firms have balked at signing and ratifying the treaty because of provisions to create an International Seabed Authority to regulate access to the ocean floor outside of states' territorial waters. Developing states had argued at the conference that the ocean was the common heritage of all people. The convention calls for technology transfers and assistance to developing states in competing for exploitation of minerals such as manganese, iron, and nickel from the deep seabed. Royalties from initial mining operations would fund a mining enterprise controlled by the International Seabed Authority. The United States sought amendments in the convention to reduce the royalties and drop the requirement for technology transfer.

The ideological arguments created by the UNCLOS process centered on the question of private property rights and incentives for entrepreneurial investments by companies versus economic equity issues for developing states. Despite the modifications won by industrial states,

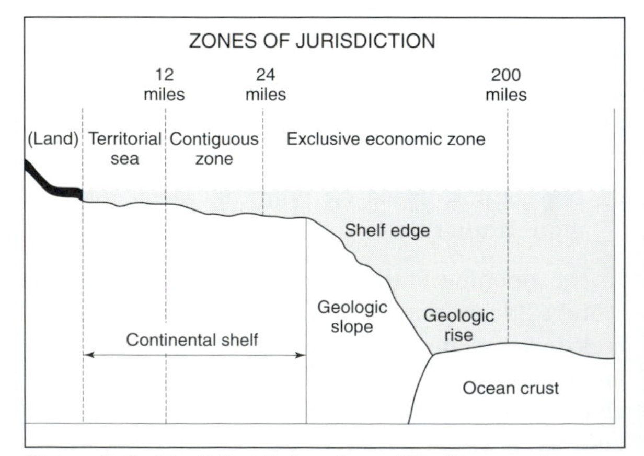

Figure 8.4 The United Nations Law of the Seas creates offshore boundaries for states

the convention does indicate a new outlook on distribution of benefits from the world's resources under international law.

The Convention on Biological Diversity, which arose from the UNCED process, ran into similar debate. The United States initially would not sign it, but did so when President Clinton took office. However, the convention has not been ratified in the US Congress, although it has received enough ratifications by other states to now be in force as international law. The section of the convention that disturbed advocates of the free market dealt with royalty payments and transfer of technology to states where biological resources are exploited. Again, the argument was put forth that companies who take the investment risk should accrue the benefits and that the provisions of the new law were contradictory to international intellectual property laws.

The UNCLOS and UNCED conventions not only demonstrate that international law has entered into the realm of resource distribution, but also that the developing states are increasingly able to defend their self-interest. However, each time that the United States or Germany and other economic giants object to an international law, the process is undermined without the support of major states. The success of international laws still depends largely on the consent of sovereign states, and when major players abstain, the game may get scuttled.

International law ultimately exists to promote the values of human society and create a regularity for state relations. The UNCLOS and UNCED conferences resulted in international conventions that promise to create a predictability for business operations and an order to world transactions. The countries that have the biggest stakes in global activities may find the laws very attractive over the long run.

BOX 8.4

Excerpts from the Final Statement of Ken Saro-Wiwa

Ken Saro-Wiwa made the following statement to the military tribunal of Nigeria who sentenced him to death in 1995 for his actions against the activities of the government and Western oil companies in the Niger Delta and against the Ogoni minority group. Nigeria was subsequently suspended from membership in the Commonwealth of Nations.

> My lord, we all stand before history. I am a man of peace, of ideas. Appalled by the denigrating poverty of my people who live on a richly endowed land; distressed by their political marginalization and economic strangulation; angered by the devastation of their land, their ultimate heritage; anxious to preserve their right to life and a decent living; and determined to usher to this country as a whole a fair and just democratic system which protects everyone and every ethnic group, and gives us all a valid claim to human civilization . . . We all stand on trial, my lord, for by our actions we have denigrated our country and jeopardized the future of our children. As we subscribe to the sub-normal and accept double standards, as we lie and cheat openly, as we protect injustice and oppression, we empty our classrooms, denigrate our hospitals, fill our stomachs with hunger and elect to make ourselves the slaves of those who ascribe to higher standards . . . I predict that the scene here will be played and replayed by future generations yet unborn. Some have already cast themselves in the role of villains, some are victims, some still have a chance to redeem themselves. The choice is for each individual.

9

Living with Paradox

It may be that the "change" in global change and the "new" in geopolitics involves the attitude of people towards the inquiry. The authors hope with this book to reinforce the notion that geopolitics is not conducted by theoreticians far removed from human realities, but is a method of approach to help each person understand his or her world. Ultimately, each of the previous chapters has dealt with human well-being: will we be protected from annihilation or exploitation? Will we be able to live in a healthy environment? Can we expect states to act on an international moral code to protect our rights as human beings? Is there even a universal code to define those rights? What is our identity with community and homeland?

In answering these questions, geography matters. We have argued that it is changing and old concepts of sovereignty and border intactness are giving way, but geography itself is not obsolete; it is evolving into a fresh way of looking at spatial relations. To believe that there is a new geopolitics, one must embrace a new geography. Therefore, let us end our initial engagement with geopolitics by returning to our issues from Chapter 1.

9.1 The role of the state

We have seen in this book that the state is still the prime actor in geopolitical events, even as its role is eroded. Yet the world has never been so full of states as today. From less than 50 states at the end of World War II, the number of states in the world today is approaching 200. If self-determination for major ethnic groups wins the day and devolutionary processes continue, the state may yet proliferate and reign supreme. But the 1992 United Nations Security Council Resolution to use military might for protecting relief supplies in Somalia may have been the most telling recent event in determining the fate of the state. In fact, we might argue that geopolitics has now entered a post-Somalian phase in which a suprasovereign organization entered a state without its consent. Somalia was disintegrating and had no functioning government, yet few would have denied at the time that it constituted a state. Its territory was intact, albeit its internal situation had deteriorated significantly. Will other states be in similar positions in the future if judged not to live up to an international code governing the constraints of sovereignty?

9.2 The concept of sovereignty

In Chapter 2 we examined the traditional definition of sovereignty and its relation to power and territorial control. The new geopolitics witnesses an erosion of sovereignty from within (due to a resurgence of nationalism and devolutionary pressures) and from without (as some control is ceded to international groups). The very concept of state sovereignty is brought into question by events such as the establishment of a Palestinian authority in the West Bank and Gaza or the impending Nunavut government creation out of the Northwest Territories of Canada in 1999. The irony of the new geopolitics is that a state might best protect its sovereignty by agreeing to give up portions of control. Will even powerful actors, such as the United States, be able to live with this? Conflicts over such agreements as the UN Law of the Seas provide some telling guideposts. Ruth Lapidoth has noted that, in order to function in today's world, states agree to transfer a considerable amount of power; indeed, those who do not become outlaws. She concludes, "sovereignty in its classic connotation of total and indivisible state power has been eroded by modern technical and economic developments and by certain rules of modern constitutional and international law" (Lapidoth 1992: 345)

9.3 The meaning of nationalism

Perhaps the most elusive concept presented in this book is that of nationalism. It is an idea that is hard to define and hard to bury. Alexander Motyl (1992) suggests that nationalism is linked to ideas of modernity born in Europe: the state, the market, democracy. It may be that as we are taken to the edge of a global culture, we retreat to the psychological safety of ethnic and national identities; thus, nationalism continues to be as strong as it ever has been in impacting international politics. In fact, Anthony Smith concludes that "the politics of ethnicity and nationalism is shot through with paradox" (1996: p. 458). Even in the modern world of states, people seek to make their own identity unique, to distinguish themselves from their neighbors. We know that in geopolitics, this process requires a grasp on territorial differentiation through the evaporating promise of the nation-state. Nations without sovereign state territory struggle to grasp their geographies: the Kurds in the Middle East, the Uighurs in Asia, Native Americans in North America.

9.4 The erosion of boundaries

We have seen that states can exist without a functioning government, with questionable moral

Plate 9.1 Egyptian–Israeli border at Sinai (photo: Kathleen Braden)

principles, with economies in disarray, but a state cannot exist without territory. To claim territorial sovereignty requires lines on a map, and those lines are fading. Our discussion in Chapter 6 on conflict, in Chapter 7 on the environment, and Chapter 8 on international law all demonstrated the permeability of borders in the new geopolitics. Yet even as boundaries cannot be protected with any embodied technology, international codes are increasingly creating a system to manage those boundaries. We demand the precision of the line; the zonal frontier has no place in the modern world – it is too imprecise, too easy to violate. Yet we continue with the illusion that military force can ensure that boundaries are intact to protect us and keep out the Otherness. Money is spent on military counters to illegal drugs, terrorism, and unwanted refugees, but the authors have argued that such attempts are akin to walling out the very air itself.

Plate 9.2 Border monument at Blaine, WA, USA to mark the longest "peaceful" (i.e. demilitarized) border in the world – that between Canada and the United States (photo: Kathleen Braden)

9.5 Scale is everything

We move from local to medium (regional and national) to global scale in geopolitical analysis. The spatial unit chosen in turn defines the work. The state scale of analysis is believed by some to be approaching its end. Kenichi Ohmae (1993), for example, has written that the nation-state has become dysfunctional as a unit for organizing human activities and suggests the rise of the region-state, linked into a global economy. Ohmae argues that the ultimate goal to emanate from capitalism is to raise the standard of living for everyone and this can best be accomplished by region-scale entities such as the east Asian one emerging for Hong Kong, Shenzen, and Guangdong Province. Would a step up in scale (to the global) or down (to the regional) indeed ensure human well-being better than the state scale has done? What scale best serves community?

9.6 Moral codes are everything else

Can it be that the proper scale for a new geopolitics must be that of community, whether perceived to be local or global? Richard Ashley (1987) warns us of a naïve assumption that modern international political community can be found in Western discourse about law and order, the ideas of Hugo Grotius and others explored in Chapter 8. But he goes on to argue that we should not assume the absence of such a community because we cannot yet recognize it. In fact, what Ashley terms the rituals of realist power politics in the capitalist world order may already hold the key to defining community. Raising a question crucial to the concepts explored in this book, Ashley asks whether late capitalist society has propelled us into a crisis because the constraints of the past imposed on statesmen (e.g. discourse on nationalism, advantages of a nuclear military) are no longer functional. Perhaps we should salvage within the existing system the elements that are positive and attempt to define what an international political community might look like.

If geography has allowed exclusion and protection through the state system of the past, how are we to thrive in a different geopolitical world? Perhaps the new geography demands living with paradox. When the reader examines what we have shown in this book, a set of ironies emerges: the state system is endangered at the very moment in history when it is most quickly proliferating; some sovereignty must be given up to keep most of it; the global culture that threatens to melt nationalism does an odd twist and may actually strengthen it as we are loathe to give up the familiarity of the "us"; borders have never been so precisely drawn and calculated, but met so little success to keep out the unwanted; and finally, the technology that has changed spatial relationships now forces human behavior to be better than it is in order to survive.

Bibliography

Agnew J A 1983 An excess of national exceptionalism: towards a new political geography of American foreign policy. *Political Geography Quarterly*, **2**: 151–166

Agnew J A 1987 *Place and Politics: the Geographical Mediation of State and Society*. Boston: Allen and Unwin

Agnew J A 1987 *The United States in the World-Economy: A Regional Geography*. Cambridge: Cambridge University Press

Agnew J A, O'Tuathail G 1988 The historiography of American geopolitics. *International Studies Quarterly*

Archer J C, Shelley F M, Leib J I 1997 The perceived geopolitical importance of the countries of the world. *Journal of Geography*, **96**: 57–62

Archer J C, Taylor P J 1981 *Section and Party: A Political Geography of American Presidential Elections from Andrew Jackson to Ronald Reagan*. New York: John Wiley

Arend A C, Beck R J 1993 *International Law and the Use of Force*. London: Routledge

Ashley R 1987 The geopolitics of geopolitical space: toward a critical social theory of international politics. *Alternatives*, **XII**: 403–434

Barone M 1990 *Our Country: The Shaping of America from Roosevelt to Reagan*. New York: Free Press

Beeley B 1990 The Turkish–Greek boundary: change and continuity. In Grundy-Warr, C (ed) *International Boundaries and Boundary Conflict Resolution: Proceedings of the 1989 IBRU Conference*. University of Durham: International Boundaries Research Unit, 29–40

Bertram G 1992 Tradeable emission permits and the control of greenhouse gases. *The Journal of Development Studies*, **28**(3): 423–446

Boutros-Ghali B 1992 Empowering the United Nations. *Foreign Affairs*, **71**(5): 89–98

Bowman I 1942 Political geography of power. *Geographical Review*, **32:** 349–352

Bowman I 1948 The geographical situation of the United States in relation to world policies. *Geographical Journal*, **112:** 129–142

Brown L R 1993 State of the world, 1993. *Worldwatch Institute*. New York: W W Norton

Brown S J, Schraub K M (ed) 1992 *Resolving Third World Conflict: Challenges for a New Era*. United States Institute of Peace Press

Bunge W 1973 The geography of human survival. *Annals of the Association of American Geographers*, **63**(3): 275–295

Burros W (ed) 1995 *Global Security Beyond 2000: Global Population Growth, Environmental Degradation, Migration, and Transnational Organized Crime*. University of Pittsburgh: Center for West European Studies

Caldwell L K 1984 *International Environmental Policy*. Durham, NC: Duke Press

Chaliand G, Rageau J 1990 *Strategic Atlas*. New York: Harper and Row

Chase-Dunn C, Hall T 1997 *Rise and Demise: Comparing World Systems*. Boulder, CO: Westview Press

Chay J (ed) 1990 *Culture in International Relations*. New York: Praeger

Chimni B S 1993 *International Law and World Order: A Critique of Contemporary Approaches*. New Delhi: Sage Publications

Cohen M, Nagel T, Scanlon T (eds) 1974 *War and Moral Responsibility*. Princeton, NJ: Princeton University Press

Cohen S 1963 *Geography and Politics in a World Divided*. New York: Random House

Cohen S B 1973 *Geography and Politics in a World Divided* 2nd edn. New York: Oxford University Press

Cohen S M 1989 *Arms and Judgment: Law, Morality,*

and the Conduct of War in the Twentieth Century. Boulder, CO: Westview Press

Cohen S M, Kliot N 1992 Place-names in Israel's ideological struggle over the administered territories. *Annals of the Association of American Geographers* **82**: (4): 653–680

Crocker C A, Hampson F O, Aall P (eds) 1996 *Managing Global Chaos: Sources of and Responses to International Conflict*. Washington DC: United States Institute of Peace Press.

CSCE (Commission on Security and Cooperation in Europe) 1991 *Report of the CSCE Meeting of Experts on National Minorities*. Washington DC: CSCE

Cutter S L 1994 Exploiting, conserving, and preserving natural resources. In Demko G, Wood W B (eds) *Reordering the World: Geopolitical Perspectives on the 21st Century*. Boulder, CO: Westview Press, 123–140

Dalby S 1990 *Creating the Second Cold War: The Discourse of Politics*. London: Pinter Publishers

De Blij H 1992 Political geography of the Post Cold War world. *Professional Geographer*, **44**(1): 16–19

Demko G, Wood W B 1994 *Reordering the World: Geopolitical Perspectives on the 21st Century*. Boulder, CO: Westview Press

De Seversky A 1950 *Air Power: Key to Survival*. New York: Simon and Schuster

Downing D 1980 *An Atlas of Territorial and Border Disputes*. London: New English Library

Elliott L M 1994 Environmental protection in the Antarctic: problems and prospects. *Environmental Politics* **3**, (2): 247–272

Fallows J 1982 *National Defense*. New York: Random House

Ferguson J 1977 *War and Peace in the World's Religions*. London: Sheldon Press

Frankel A, Frankel B 1987 Israel's nuclear ambiguity. *Bulletin of the Atomic Scientists*, March: 15–19

Fussell P 1989 *Wartime: Understanding and Behavior in the Second World War*. New York and Oxford: Oxford University Press.

Garreau J 1982 *The Nine Nations of North America*. New York: Avon Books

Geisel A 1984 *The Butter Battle Book*. New York: Random House

Glassner M 1993 *Political Geography*. John Wiley

Glassner M I, De Blij H J 1989 *Systematic Political Geography*, 4th edn. New York: John Wiley

Goldstein J S 1988 *Long Cycles: Prosperity and War in the Modern Age*. New Haven: Yale University Press

Greene O, Percival I, Ridge I 1985 *Nuclear Winter*. Cambridge: Polity Press

Greenfeld L 1992 *Nationalism: Five Roads to Modernity*. Harvard University Press

Grundy-Warr C (ed) 1990 *International Boundaries and Boundary Conflict Resolution: Proceedings of the 1989 IBRU Conference*. University of Durham: International Boundaries Research Unit

Gupta J 1995 The global environment in its North–South context. *Environmental Politics*, **4**(1): 19–43

Hardin G 1977 The tragedy of the commons. In Hardin G J, Baden J (eds) *Managing the Commons*. San Francisco: W H Freeman

Hartmann F 1982 *The Conservation of Enemies: A Study in Enmity*. Westport, CO: Greenwood Press

Held V, Morgenbesser S, Nagel T 1974 *Philosophy, Morality, and International Affairs*. New York: Oxford University Press.

Hewitt K 1983 Place annihilation: area bombing and the fate of urban places. *Annals of the Association of American Geographers*, **73**(2): 257–284

Higgins R 1994 *Problems and Process: International Law and How We Use It*. Oxford: Clarendon Press

Hughes B B 1993 *International Futures: Choices in the Creation of a New World Order*. Boulder, CO: Westview Press

Huntington S 1993 The clash of civilizations? *Foreign Affairs*, **72**(3): 22–49

Hurrell A, Kingsbury B (eds) 1992 *The International Politics of the Environment*. Oxford: Clarendon Press

International Bank for Reconstruction and Development 1992 *World Development Report 1992: Development and the Environment*. Oxford University Press

Johnston R J, Taylor P J 1986 *A World in Crisis?* Oxford and New York: Basil Blackwell

Johnston R J, Taylor P J, Watts M J 1995 *Geographies of Global Change: Remapping the World in the Late Twentieth Century*. Oxford: Blackwell

Jordan A 1994 Paying the incremental cost of global environmental protection: the evolving role of GEF. *Environment*, **36**(4): 12–20, 31–36

Kamieniecki S 1993 *Environmental Politics in the International Arena*. State University of New York Press

Kaplan R 1994 The coming anarchy. *The Atlantic Monthly*, **273**(2): 44–76

Keen S 1986 *Faces of the Enemy: Reflections of the Hostile Imagination*. San Francisco: Harper & Row

Kellert S R 1996 *The Value of Life: Biological*

Diversity and Human Society. Washington DC: Island Press
Knight D 1982 Identity and territory: geographical perspectives on nationalism and regionalism. *Annals of the Association of American Geographers*, **72:** 514–531
LaCroix W L 1988 *War and International Ethics: Tradition and Today*. Lanham, MD: University Press of America
Laferriere E 1994 Environmentalism and the global divide. *Environmental Politics*, **3**(1): 91–113
Lapidoth R 1992 Sovereignty in transition. *Journal of International Affairs*, **45**(2): 325–345
Larson B, Shah A 1994 Global tradeable carbon permits, participation incentives, and transfers. *Oxford Economic Papers*, **46:** 841–856
Luard E (ed) 1970 *The International Regulation of Frontier Disputes*. New York: Praeger
Lund M S 1996 *Preventing Violent Conflicts: A Strategy for Preventative Diplomacy*. Washington DC: United States Institute of Peace Press
Mackinder H J 1904 The geographical pivot of history. *Geographical Journal*, **23:** 421–442
Mackinder H J 1919 *Democratic Ideals and Reality: A Study in the Politics of Reconstruction*. London: Constable
Mackinder H J 1943 The round world and the winning of the peace. *Foreign Affairs*, **21:** 595–605
MacLeish A 1980 The conquest of America. *The Atlantic Monthly*, March, 35–42. Reprinted from 1949
Mahan A T 1890 *The Influence of Sea Power Upon History*. Boston: Little, Brown & Co
McNeill W 1982 *The Pursuit of Power: Technology, Armed Force, and Society Since* AD *1000*. Chicago: University of Chicago Press
Meinig D W 1986 *The Shaping of America: A Geopolitical Perspective on 500 Years of History*. Yale University Press
Mikesell M, Murphy A 1991 A framework for comparative study of minority-group aspirations. *Annals of the Association of American Geographers*, **81**(4): 581–604
Modelski G 1987 *Long Cycles of World Politics*. London: MacMillan
Mofson P 1994 Global ecopolitics. In Demko G, Wood W B (eds) *Reordering the World: Geopolitical Perspectives on the 21st Century*. Boulder, CO: Westview Press, 167–178
Morrill R 1994 *Annals of the Association of American Geographers*, **74**(1): 7–8
Motyl A 1992 The modernity of nationalism: nations, states and nation-states in the contemporary world. *Journal of International Affairs*, **45**(2): 308–323
Murphy A 1990 Historical justifications for territorial claims. *Annals of the Association of American Geographers*, **80**(4): 531–548
Naess A 1983 The deep ecology movement: some philosophical aspects. In Zimmerman M (ed.) *Environmental Philosophy*. Englewood Cliffs: Prentice Hall
Nardin T, Mapel D R (eds) 1992 *Traditions of International Ethics*. Cambridge: Cambridge University Press
National Military Strategy 1992 Joint Chiefs of Staff, United States Armed Forces
Nietschmann B 1994 The fourth world: nations versus states. In Demko G, Wood W (eds) *Reordering the World: Geopolitical Perspectives on the Twenty-first Century*. Boulder, CO: Westview Press
Nijman J *et al.* (eds) 1992 The political geography of the post Cold War world. *The Professional Geographer*, **44**(1): 1–29
Novak M 1983 *Moral Clarity in the Nuclear Age*. Nashville, TN: Thomas Nelson
Ó Tuathail G, Luke T 1994 Present at the (dis)integration: deterritorialization and reterritorialization in the New Wor(l)d Order. *Annals of the Association of American Geographers*, **84**(3): 381–398
O'Loughlin J 1986 Spatial models of international conflicts: extending current theories of war behavior. *Annals of the Association of American Geographers*, **76**(1): 63–80
O'Loughlin J, van der Wusten H 1986 Geography, war and peace: notes for a contribution to a revived political geography. *Progress in Human Geography*, **10**: 484–510
Ohmae K 1993 The rise of the region state. *Foreign Affairs*, **72**(2): 78–96
Orwell G 1945 Notes on nationalism. *Collected Essays: Journalism and Letters of George Orwell*, vol 3. New York: Harcourt Brace & Jovanovich
O'Sullivan P 1986 *Geopolitics*. London: Croom Helm
Parker G 1985 *Western Geopolitical Thought in the Twentieth Century*. New York: St Martin's Press
Parsons J 1985 On bioregionalism and watershed consciousness. *The Professional Geographer*, **37**(1): 1–6
Paterson J H 1987 German geopolitics reassessed. *Political Geography Quarterly*, **8:** 107–114
Pepper D, Jenkins A 1985 *The Geography of Peace and War*. Oxford and New York: Basil Blackwell
Pfaff W 1993 *The Wrath of Nations: Civilization and*

the Furies of Nationalism. New York: Simon & Schuster

Phillips A, Rosas A 1995 *Universal Minority Rights*. Turku/Abo: Abo Akademi University Institute for Human Rights, Finland

Plumwood V 1995 Has democracy failed ecology? An ecofeminist perspective. *Environmental Politics*, **4**(4): 134–168

Porter G, Brown J W 1991 *Global Environmental Politics*. Boulder, CO: Westview

Powell C 1992 *The National Military Strategy 1992*. Washington DC: Joint Chiefs of Staff

Raine L P, Cillufo F J (eds) 1994 *Global Organized Crime: The New Empire of Evil*. Washington DC: Center for Strategic and International Studies

Ramsbotham O, Woodhouse T 1996 *Humanitarian Intervention in Contemporary Conflict: A Reconceptualization*. Cambridge, UK: Polity Press

Raustiala K, Victor D G 1996 The future of the convention on biological diversity. *Environment*, **38**(4): 16–19, 37–45

Righter R 1995 *Utopia Lost: The United Nations and the World Order*. New York: Twentieth Century Fund Press

Sack R 1983 Human territoriality: a theory. *Annals of the Association of American Geographers*, **73**(1): 55–74

Schachter O 1991 The emergence of international environmental law. *Journal of International Affairs*, **44**(2): 457–493

Simmons I G 1993 *Interpreting Nature: Cultural Constructions of the Environment*. London and New York: Routledge

Smith A D 1996 Culture, community and territory: the politics of ethnicity and nationalism. *International Affairs*, **72**(3): 445–458

Somers E 1987 Environmental hazards show no respect for national boundaries. *Environment*, **29**(5): 7–9, 31–33

Spykman N J 1942 *America's Strategy in World Politics: The United States and the Balance of Power*. London: Harcourt Brace

Taylor P J 1990 *Political Geography*, 2nd edn. London: Longman

Taylor P 1992 *Britain and the Cold War: Nineteen Forty-Five As Geopolitical Transition*. London and New York: Guilford Publications

Thompson J 1995 Toward a green world order: environment and world politics. *Environmental Politics*, **4**(4): 31–48

Thompson S L, Aleksander V V, Stenchikov G L, Schneider S H, Covey C, Chervin R M 1984 Global climatic consequences of nuclear war: simulation with three-dimensional model. *Ambio*, **13**: 263–243

Thrift N 1992 Muddling through: world orders and globalization. *The Professional Geographer*, **44**(1): 3–7.

Trolldalen J M 1992 *International Environmental Conflict Resolution: The Role of the United Nations*. Oslo: World Foundation for Environment and Development

Trubowitz P 1998 *Defining the National Interest: Conflict and Change in American Foreign Policy*. Chicago: University of Chicago Press

Tucker M E, Grim J 1993 *Worldviews and Ecology*. Lewisburgh, PA: Bucknell University Press

United States Catholic Conference 1983 *The Challenge of Peace*. US Bishops' Pastoral Letter

Vernadsky V I 1998 *The Biosphere* (tranlsated by D Langmuir). New York: Copernicus

Volkan V 1988 *The Need to Have Enemies and Allies*. Northvale, NJ: Jason Aronson

Wakin M (ed) 1986 *War, Morality, and the Military Profession*, 2nd edn. Boulder, CO: Westview Press

Wallerstein I 1991 *Geopolitics and Geoculture: Essays on the Changing World System*. Cambridge University Press

Wapner P 1995 The state and environmental challenges: a critical exploration of alternatives to the state-system. *Environmental Politics*, **4**(1): 44–69

Ward M D (ed) 1992 *The New Geopolitics*. Philadelphia: Gordon and Breach

Whittlesey D 1939 *The Earth and the State*. New York: Henry Holt and Co

World Resources Institute 1994 *World Resources 1994–95: A Guide to the Global Environment*. New York: Oxford University Press

Zelinsky W 1989 *Nation Into States: The Shifting Symbolic Foundations of American Nationalism*. Chapel Hill, NC: University of North Carolina Press

Zimmerman M 1994 *Contesting Earth's Future: Radical Ecology and Postmodernity*. Berkeley, CA: University of California Press

Index